Dimensional Analysis Beyond the Pi Theorem

Bahman Zohuri

Dimensional Analysis Beyond the Pi Theorem

Springer

Bahman Zohuri
Galaxy Advanced Engineering, Inc.
San Mateo, California, USA

ISBN 978-3-319-45725-3   ISBN 978-3-319-45726-0 (eBook)
DOI 10.1007/978-3-319-45726-0

Library of Congress Control Number: 2016949733

Printed on acid-free paper

This Springer imprint is published by Springer Nature
The registered company is Springer International Publishing AG Switzerland
The registered company address is: Gewerbestrasse 11, 6330 Cham, Switzerland

To one of my best friends Bill Kemp, his wife Linda, and daughter Jennifer

About the Author

Dr. Bahman Zohuri currently works for Galaxy Advanced Engineering, Inc., a consulting firm that he started in 1991 when he left both the semiconductor and defense industries after many years working as a chief scientist. After graduating from the University of Illinois in the field of physics and applied mathematics, he then went to the University of New Mexico, where he studied nuclear engineering and mechanical engineering. He joined Westinghouse Electric Corporation, where he performed thermal hydraulic analysis and studied natural circulation in an inherent shutdown and heat removal system (ISHRS) in the core of a liquid metal fast breeder reactor (LMFBR) as a secondary fully inherent shutdown system for secondary loop heat exchange. All these designs were used in nuclear safety and reliability engineering for a self-actuated shutdown system. He designed a mercury heat pipe and electromagnetic pumps for large pool concepts of an LMFBR for heat rejection purposes for this reactor around 1978, when he received a patent for it. He was subsequently transferred to the defense division of Westinghouse, where he oversaw dynamic analysis and methods of launching and controlling MX missiles from canisters. The results were applied to MX launch seal performance and muzzle blast phenomena analysis (i.e., missile vibration and hydrodynamic shock formation). Dr. Zohuri was also involved in analytical calculations and computations in the study of nonlinear ion waves in rarefying plasma. The results were applied to the propagation of so-called soliton waves and the resulting charge collector traces in the rarefaction characterization of the corona of laser-irradiated target pellets. As part of his graduate research work at Argonne National Laboratory, he performed computations and programming of multi-exchange integrals in surface physics and solid-state physics. He earned various patents in areas such as diffusion processes and diffusion furnace design while working as a senior process engineer at various semiconductor companies, such as Intel Corp., Varian Medical Systems, and National Semiconductor Corporation. He later joined Lockheed Martin Missile and Aerospace Corporation as senior chief scientist and oversaw research and development (R&D) and the study of the vulnerability, survivability, and both radiation and laser hardening of different components of the Strategic Defense Initiative, known as Star Wars.

This included payloads (i.e., IR sensor) for the Defense Support Program, the Boost Surveillance and Tracking System, and Space Surveillance and Tracking Satellite against laser and nuclear threats. While at Lockheed Martin, he also performed analyses of laser beam characteristics and nuclear radiation interactions with materials, transient radiation effects in electronics, electromagnetic pulses, system-generated electromagnetic pulses, single-event upset, blast, and thermomechanical, hardness assurance, maintenance, and device technology.

He spent several years as a consultant at Galaxy Advanced Engineering serving Sandia National Laboratories, where he supported the development of operational hazard assessments for the Air Force Safety Center in collaboration with other researchers and third parties. Ultimately, the results were included in Air Force Instructions issued specifically for directed energy weapons operational safety. He completed the first version of a comprehensive library of detailed laser tools for airborne lasers, advanced tactical lasers, tactical high-energy lasers, and mobile/tactical high-energy lasers.

He also oversaw SDI computer programs, in connection with Battle Management C^3I and artificial intelligence, and autonomous systems. He is the author of several publications and holds several patents, such as for a laser-activated radioactive decay and results of a through-bulkhead initiator. He has published the following works: *Heat Pipe Design and Technology: A Practical Approach* (CRC Press); *Dimensional Analysis and Self-Similarity Methods for Engineers and Scientists* (Springer); *High Energy Laser (HEL): Tomorrow's Weapon in Directed Energy Weapons Volume I* (Trafford Publishing Company); and recently the book on the subjects directed energy weapons and physics of high-energy laser with Springer. He has other books with Springer Publishing Company: *Thermodynamics in Nuclear Power Plant Systems* and *Thermal-Hydraulic Analysis of Nuclear Reactors*.

Preface

In physics and science, dimensional analysis is a tool to find or check relations among physical quantities by using their dimensions. The dimension of a physical quantity is the combination of the basic physical dimensions (usually mass, length, time, electric charge, and temperature) which describe it. For example, speed has the dimension length/time and may be measured in meters per second, miles per hour, or other units. Dimensional analysis is necessary because a physical law must be independent of the units used to measure the physical variables in order to be general for all cases.

Dimensional analysis is routinely used to check the plausibility of derived equations and computations as well as forming reasonable hypotheses about complex physical situations that can be tested by experiment or by more developed theories of the phenomena, which allow categorizing the types of physical quantities. In this case, units are based on their relations or dependence on other units or dimensions, if any.

Isaac Newton (1686) who referred to it as the "Great Principle of Similitude" understood the basic principle of dimensional analysis. The nineteenth-century French mathematician Joseph Fourier made important contributions based on the idea that physical laws like $F = MA$ should be independent of the units employed to measure the physical variables. This led to the conclusion that meaningful laws must be homogeneous equations in their various units of measurement, a result that was eventually formalized by Edgar Buckingham with the π (pi) theorem. This theorem describes how every physically meaningful equation involving n variables can be equivalently rewritten as an equation of $n–m$ dimensionless parameters, where m is the number of fundamental dimensions used. Furthermore, and most importantly, it provides a method for computing these dimensionless parameters from the given variables.

A dimensional equation can have the dimensions reduced or eliminated through nondimensionalization, which begins with dimensional analysis and involves scaling quantities by characteristic units of a system or natural units of nature.

The similarity method is one of the standard methods for obtaining exact solutions of partial differential equations (PDEs), in particular nonlinear forms. The number of independent variables in a PDE is reduced one by one to make use of appropriate combinations of the original independent variables as new independent variables, called "similarity variables."

In some cases, dimensional analysis does not provide an adequate approach to establish a solution of a certain eigenvalue problem in nonlinear form which gives rise to the need to discuss similarity method as another approach. In particular, simple cases deal with reduction of a partial differential equation to an ordinary differential equation in an ordinary way that we have learned in any classical text of the same type. In more scenarios that are complex, dealing with boundary value problem for a system of ordinary equations with conditions at different ends of an infinite interval requires to construct a self-similar solution that is a more efficient way of solving such complex boundary value problem for the system of ordinary equations directly. In a specific instance, the passage of the solution into a self-similar intermediate asymptotic prevents a return to the partial differential equations; indeed, in many cases, the self-similarity of intermediate asymptotic can be established and the form of self-similar intermediate asymptotic obtained from dimensional considerations.

For the subject of this book *Dimensional Analysis Beyond the Pi Theorem*, we are looking beyond just the simple pi theorem. Although the dimensional analysis and physical similarity are well-understood subjects, the general concepts of dynamical similarity are explained in this book. Our exposition is essentially different from those available in the literature, although it follows in its general ideas that are known as the pi theorem and there are many excellent books by the different authors that are published, which one can refer to. However, dimensional analysis goes way beyond the pi theorem, which is also known as Buckingham's pi theorem. Many techniques via self-similar solutions can bound these solutions to problems that seem to be intractable.

The human partner in the interaction of a man and a computer often turns out to be the weak spot in the relationship. The problem of formulating rules and extracting ideas from vast masses of computational or experimental results remains a matter for our brains, our minds. This problem is closely connected with the recognition of patterns. The word "obvious" has two meanings, not only something easily and clearly understood but also something immediately evident to our eyes. The identification of forms and the search for invariant relations constitute the foundation of pattern recognition; thus, we identify the similarity of large and small triangles.

A time-developing phenomenon is called self-similarity if the spatial distributions of its properties at various different moments of time can be obtained from one another by a similarity transformation, and the fact that we identify one of the independent variables of dimension with time is nothing new from the subject of dimensional analysis point of view. However, this is where the boundary of dimensional analysis goes beyond the pi theorem and steps into a new arena that is known as self-similarity, which has always represented progress for researchers.

In recent years, there has been a surge of interest in self-similar solutions of the first and second kind. Such solutions are not newly discovered; they had been identified and in fact so named by Zel'dovich, a famous Russian mathematician, in 1956, in the context of a variety of problems, such as shock waves in gas dynamics and filtration through elastoplastic materials.

Self-similarity has simplified computations and the representation of the properties of phenomena under investigation. It handles experimental data, reduces what would be a random cloud of empirical points to lie on a single curve or surface, and constructs procedure that is known to us as self-similar, wherein variables can be chosen in some special way.

The self-similarity of the solutions of partial differential equations either in linear or nonlinear form has allowed their reduction to ordinary differential equations, which often simplifies the investigation. Therefore, with the help of self-similar solutions, researchers and scientists have attempted to envisage the characteristic properties of new phenomena.

Nonlinearity plays a major role in the understanding of most physical, chemical, biological, and engineering sciences. Nonlinear problems fascinate scientists and engineers, but often elude exact treatment. However elusive they may be, the solutions do exist—if only one perseveres in seeking them out.

Although the book does not provide any exercises at the end of each chapter, throughout the book, numerous examples are provided for the appropriate chapter and sections. Thus, the reader will have ample practical examples of dimensional problems instead of facing a cut and dry abstract approach as existing books of this subject follow.

Note that I have tried to make this book stand alone, yet give enough background if the reader has no background on dimensional analysis or the pi theorem at all. Consequently, I have kept Chaps. 1 and 4 of this book similar to what I have published before, which is a book with Springer Publishing Company, under the title *Dimensional Analysis and Self-Similarity Methods for Engineers and Scientists* in 2015.

San Mateo, CA Bahman Zohuri

Acknowledgments

I am indebted to the many people who aided me, encouraged me, and supported me beyond my expectations. Some are not around to see the results of their encouragement in the production of this book, yet I hope they know of my deepest appreciations. I especially want to thank my friend Bill Kemp, to whom I am deeply indebted. He has continuously given his support without hesitation and has always kept me going in the right direction.

Above all, I offer very special thanks to my late mother and father and to my children, in particular, my son Sasha. They have provided constant interest and encouragement, without which this book would not have been written. Their patience with my many absences from home and long hours in front of the computer to prepare the manuscript is especially appreciated.

About This Document

The purpose of this document is to describe dimensional analysis, similarity, and modeling methods.

Contents

Chapter 1
Principles of the Dimensional Analysis

Nearly all scientists at conjunction with simplifying a differential equation have probably used dimensional analysis. Dimensional analysis (also called the factor-label method or the unit factor method) is an approach to problem that uses the fact that one can multiply any number or expression without changing its value. This is a useful technique. However, the reader should take care to understand that chemistry is not simply a mathematics problem. In every physical problem, the result must match the real world.

1.1 Introduction

Dimensional analysis is a method by which we deduce information about a phenomenon from the single premise that a phenomenon can be described by a dimensionally consistent equation of certain variables. The generality of the method is both its strength and its weakness. The result of a dimensional analysis of a problem is to reduce the number of variables in the problem, thereby gathering sufficient information from only a few experiments.

Dimensional analysis treats the general forms of equations that describe natural phenomena, and its application is abounded in nearly all fields of engineering, particularly in fluid mechanics and in heat transfer theory.

The application of dimensional analysis to any particular phenomenon is based on the assumption that certain named variables are the independent variables of the problem and that aside from the dependent variable, all others are redundant or irrelevant. This initial step—the naming of the variables—often requires and sometimes brings a philosophic insight into the natural phenomena that is being investigated.

The first step in modeling any physical phenomena is the identification of the relevant variables and then relating these variables via known physical laws. For sufficiently simple phenomena, we can usually construct a quantitative relationship

B. Zohuri, *Dimensional Analysis Beyond the Pi Theorem*,
DOI 10.1007/978-3-319-45726-0_1

among these variables from first principles; however, many complex phenomena (which often occur in engineering applications) such an ab initio theory are often difficult, if not impossible. In these situations, modeling methods are indispensable, and one of the most powerful modeling methods is dimensional analysis. We have probably encountered dimensional analysis in our previous physics courses when we were admonished to "check our units" to ensure that the left- and right-hand sides of an equation had the same units (e.g., so that our calculation of a force had the units of kg m/s^2). In a sense, this is why there is need for dimensional analysis, although checking units is certainly the most trivial example of dimensional analysis. Here we will use dimensional analysis to actually *solve* problems or at least infer valuable information about the solution.

According to Professor G. I. Barenblatt of University California at Berkeley, "many of those who have taught dimensional analysis (or have merely thought about how it should be taught) have realized that it has suffered an unfortunate fate. In fact, the idea on which dimensional analysis is based on, is very simple, and can be, understood by everybody: physical laws do not depend on arbitrarily chosen basic units of measurement. An important conclusion can be drawn from this simple idea, using a simple argument: the functions that express physical laws must possess a certain fundamental property, which in mathematics is called generalized homogeneity or symmetry. This property allows the number of arguments in these functions to be reduced, thereby making it simpler to obtain them (by calculating them or determining them experimentally). This is, in fact, the entire content of dimensional analysis—there is nothing more to it."

The basic idea is the following: *physical laws do not depend upon arbitrariness in the choice of the basic units of measurement*. In other words, Newton's second law, $F = ma$, is true whether we choose to measure mass in kilograms, acceleration in meters per second squared, and force in Newton's, or whether we measure mass in slugs, acceleration in feet per second squared, and force in pounds. As a concrete example, consider the angular frequency of small oscillations of a point pendulum in small angle oscillation with length l and mass m:

$$\omega = \sqrt{\frac{g}{l}} \tag{1.1}$$

where g is the acceleration due to gravity, which is 9.8 m/s^2 on earth in the SI system of units. To derive Eq. 1.1, one usually needs to solve the differential equation which results from applying Newton's second law to the pendulum (do it!). See Appendices A and B for the analysis. Let us instead deduce (Eq. 1.1) from dimensional considerations alone. What can ω depend upon? It is reasonable to assume that the relevant variables are m, l, and g (it is hard to imagine others, at least for a point pendulum). Now suppose that we change the system of units so that the unit of mass is changed by a factor of M, the unit of length is changed by a factor of L, and the unit of time is changed by a factor of T. With this change of units, the units of frequency will change by a factor of g, the units of velocity will change by a factor of LT^{-1}, and the units of acceleration by a factor of LT^{-2}. Therefore, the units of the quantity g/l will change by T^{-2}, and those of $(g/l)^{1/2}$ will change by T^{-1}. Consequently, the ratio

$$\Pi = \frac{\omega}{\sqrt{g/l}} \tag{1.2}$$

is invariant under a change of units; Π is called a *dimensionless number*. Since it does not depend upon the variables (m, g, l), it is in fact a constant. Therefore, from dimensional considerations alone we find that

$$\omega = \text{constant} \cdot \sqrt{g/l} \tag{1.3}$$

A few comments are in order:

1. The frequency is independent of the mass of the pendulum bob, a somewhat surprising conclusion to the uninitiated.
2. The constant cannot be determined from dimensional analysis alone.

These results are typical of dimensional analysis—we uncover often-unexpected relations among the variables, while at the same time we fail to pin down numerical constants. Indeed, to fix the numerical constants we need a real *theory* of the phenomena in question, which goes beyond simple dimensional considerations.

Unites is a quantitative statement about an objective magnitude and is necessarily composed of two parts or factors: a number and a statement of the unit of measurement. The number is the mathematical ratio of the magnitude to that of the specific unit. Similarly, the ultimate end of all applied mathematics is the numerical evaluation, by the working of an arithmetical sum, of the magnitude of some physical quantity, which is inferred from the known magnitudes of others. Any physical quantity can be completely defined by a number and any arbitrarily valued unit, provided that, the unit is exactly specified and relevant to the physical system. A collection of units for the measurement of physical quantities is known as a system of units, and, in such a system, the various units may be either arbitrarily defined, or they may be made to depend in a simple way on other units. Per Prof. Barenblatt, The units for measuring physical quantities are divided into two categories: *fundamental units and derived units*. This means the following:

A class of phenomena (e.g., mechanics, i.e., the motion and equilibrium of bodies) is singled out for study. Certain quantities are listed, and standard reference values for these quantities either natural or artificial are adopted as fundamental units; there is a certain amount of arbitrariness here. For example, when describing mechanical phenomena, we may adopt mass, length, and time standards as the fundamental units, even though it is also possible to adopt force, length, and time standards. However, these standards are insufficient for the description of, for example, heat transfer due to tempera. Additional standards also become necessary when studying electromagnetic phenomena, etc.

A set of fundamental units, which is sufficient for measuring the properties of the class of phenomena under consideration, is called a *system of units*. Until recently, the centimeter–gram–second (CGS) system, in which units for mass, length, and time are used as the basic units and 1 gram (g) is adopted as the unit of mass, 1 centimeter (cm) is adopted as the unit of length, and 1 second (s) is adopted as the unit of time, has customarily been used. However, a system of units need not be *minimal* [1]. For example, one can use a system of units in which the unit of length is 1 cm, the unit of time is 1 s, and the unit of velocity is 1 knot (approximately 50 cm/s). However, in the case of this system, the velocity will not be numerically equal to the ratio of the distance traversed to the magnitude of the time interval in which this distance was traversed.

Classes of systems of units: In addition to the CGS system, there is a second system, in which a standard length of 1 km (=10^5 cm) is used as the unit of length. A standard mass of 1 metric ton (=10^6 g) is used as the unit mass, and finally, a standard time interval of 1 h (=3600 s) is used as the unit of time. These two systems of units have the following property in common:

- Standard quantities of the same physical nature (mass, length, and time) are used fundamental units. Consequently, we say that these systems belong to the same *class*. To generalize, a set of systems of units that differ only in the magnitude (but not in the physical nature) of the fundamental units is called a *class of systems of units*.
- The system just mentioned and the CGS systems are members of the class in which standard lengths, masses, and the times are used as the fundamental units.

The corresponding units for an arbitrary system in this class are as follows:

$$\begin{aligned} &\text{Unit of length} = \text{cm} \quad L \\ &\text{Unit of mass} \;\, = \text{g} \quad\;\; M \\ &\text{Unit of time} \;\;\, = \text{s} \quad\;\; T \end{aligned}$$

where L, M, and T *are abstract positive numbers* that indicate the factors by which the fundamental units of length, mass, and time decrease in passing from the original system (in this case, the CGS system) to another system in the same class. This class is called the *LMT* class.

Note: The designation of a class of system of units is obtained by writing down, in consecutive order, the symbols for the quantities whose units are adopted as the fundamental units. These symbols simultaneously denote the factor by which the corresponding fundamental unit decreases upon passage from the original system to another system in the same class.

The SI (MKS) system has recently come into widespread use. This system, in which 1 m (=100 cm) is adopted as the unit of length, 1 kg (=1000 g) is adopted as the unit of mass, and 1 s is adopted as the unit of time, also belongs to the *LMT* class. Thus, when passing from the original system to the SI system, $M = 0.001$, $L = 0.01$, and $T = 1$.

Systems in the *LFT* class, where units for length, force, and time are chosen as the fundamental units, are also frequently used; the fundamental units for this class are as follows:

$$\begin{aligned} &\text{Unit of length} = \text{cm} \quad L \\ &\text{Unit of force} \;\, = \text{kgf} \quad F \\ &\text{Unit of time} \quad = \text{s} \quad\;\; T \end{aligned}$$

The unit of force in the original system, the kilogram force (kgf), is the force that imparts an acceleration of 9.80665 m/s^2 to a mass equal to that of the standard kilogram.

Note: a change in the magnitudes of the fundamental units in the original system of units does not change the class of systems of units.

For example, the classes where the units of length, mass, and time are given by

$$N(\chi_i, \tau)_p = N(\chi_i, \tau)_m u_1^\alpha u_2^\beta \ldots u_n^\eta$$

$$\left(n_{u1}^\alpha n_{u2}^\beta \ldots n_{un}^\eta\right)_p = \left(n_{u1}^\alpha n_{u2}^\beta \ldots n_{un}^\eta\right)_m$$

$$N_{\text{op}} = N_{\text{om}}$$

$$n_u = u/u_0$$

$$\frac{\left(u_1^\alpha u_2^\beta \ldots u_n^\eta\right)_p}{\left(n_{10}^\alpha n_{20}^\beta \ldots n_{20}^\eta\right)_p} = \frac{\left(n_1^\alpha n_2^\beta \ldots n_n^\eta\right)_m}{\left(n_{10}^\alpha n_{20}^\beta \ldots n_{n0}^\eta\right)_m}$$

$$k_u = \frac{u(\chi_i, \tau)_p}{u(\chi_i, \tau)_m} = \frac{u_{\text{op}}}{u_{\text{om}}} = \text{constant}$$

are the same as that defined in *LMT*. The only difference is that the numbers *L*, *M*, and *T* for a given system of units (e.g., the SI system) will be different in the two representations of *LMT* class in the second representation; we obviously have $L = 1$, $M = 1$, and $T = 3600$.

Units of Force and Mass

If Newton's second law of motion, $F = ma$, is applied to a freely falling body, the force *F* is the weight *W*, and the acceleration "*a*" is the acceleration of gravity of *g*. Consequently, $W = mg$ or $m = W/g$. By setting $m = 1$ in this equation, we perceive the following general rule: for consistency with $F = ma$, the weight of a unit mass must be exactly *g* units of force. The conventional systems of measurement conform to this rule. Six common systems of measurement are generally used.

(continued)

(continued)

CGS System—The letters CGS denote, respectively, "centimeter," "gram," and "second." The gram is generally regarded as a unit of mass. This is the thousandth part of a kilogram, the latter mass being legally defined as the mass of a platinum cylinder that is deposited at the International Bureau of Weights and Measures at Sevres, France. The kilogram is very nearly the mass of a liter of water at 4 °C. A balance-type scale accurately measures masses.

For consistency with Newton's second law of motion, the unit of force in the CGS system is defined to be that force which will give a gram mass an acceleration of 1 cm/s. This unit of force is called a "dyne." Since the standard value of g is 980.665 cm/s, the equation $W = mg$ shows that the weight of a gram mass on the earth is approximately 981 dynes. In the CGS system, the unit of work is a "dyne centimeter." This unit is called an "erg."

MKS Mass System—The letters MKS denote, respectively, "meter," "kilogram," and "second." In the MKS mass system, the kilogram is regarded as a unit of mass. For consistency with $F = \mathrm{ma}$, the unit of force is that force which will give a kilogram mass an acceleration of 1 m/s ^{2}. This unit of force is called a Newton. Evidently, a Newton is one hundred thousand dynes. This is approximately 0.225 lb. The unit of work in the MKS mass system is a Newton meter. This unit is called a joule. The joule is ten million ergs. The watt is a unit of power that is defined to be one joule per second.

MKS Force System—In the MKS force system, the kilogram is regarded as a unit of force, rather than a unit of mass. The kilogram force is defined to be the weight of a kilogram mass under standard gravitational attraction. Consequently, a kilogram force is 980,665 dynes. For consistency with the equation, $F = ma$, the unit of mass in the MKS force system is a "kilogram second squared per meter" (kg s^2/m). This unit of mass has not received a special name. The equation $W = mg$ shows that 1 kg s^2/m is a mass that weighs approximately 9.81 kg force on the earth. The MKS force system is extensively used in engineering practice in continental Europe.

British Mass System—In the British mass system, the pound is regarded as a unit of mass. The pound mass is legally defined to be 0.4536 kg mass. For consistency with the equation, $F = ma$, the unit of force is defined to be that force which will give a pound mass an acceleration of one foot per second squared. This unit of force is called a pound. Since the standard acceleration of gravity is 32.174 ft/s^2, the equation, $W = mg$, shows that the weight of a pound mass on the earth is approximately 32.2 lb. The British mass system is frequently used in British technical writings.

American Engineering System (US Customary Units)—Engineers in the United States usually regard the pound as a unit of force, namely,

(continued)

(continued)

0.4536 kg force. For consistency with the equation, $F = ma$, the unit of mass is then a "pound second square per foot" (lb s^2/ft). This unit of mass is called a "slug." The equation $W = mg$ shows that the slug is a mass that weighs approximately 32.2 lb on the Earth. The American engineering system does not logically exclude the concept of a "pound of matter." A pound of matter may be defined to be the quantity of matter that weighs one pound on a spring scale. This is not an invariable quantity of matter, since it depends on the local acceleration of gravity.

1.2 Dimensional Analysis and Scaling Concept

Scaling is the branch of measurement that involves the construction of an instrument that associates qualitative constructs with quantitative metric units, and the term describes a very simple situation. S.S. Stevens came up with the simplest and most straightforward definition of scaling. He said:

Scaling is the assignment of objects to numbers according to a rule.

However, what does that mean?

Most physical magnitudes characterizing nanoscale systems differ enormously from those familiar in macroscale systems. Estimate some of these magnitudes, however, by applying scaling laws to the values for macroscale systems. There are many different scaling laws. At one extreme, there are simple scaling laws that are easy to learn, easy to use, and very useful in everyday life. This has been true since day 1 of modern science. Galileo presented several important scaling results in 1638 [2].

The existence of a power-law relationship between certain variables y and x

$$y = Ax^{\alpha} \tag{1.4}$$

where A and α are constant values. This type of relationship often can be seen in the mathematical modeling of various phenomena, not only in mechanical engineering and physics but also in other science fields such as biology, economics, and other engineering discipline.

Distribution of power law is unique and has certain interesting features and graphically can be presented as a log–log scales as a straight line. This can methodically be shown, if we take the base 10 of logarithm of Eq. 1.4 as follows:

$$\begin{cases} \log(y) = \log(Ax^{\alpha}) \\ \log(y) = \log A + \log x^{\alpha} \\ \text{Assume } \log A = B \text{ Then} \\ \log y = B + \alpha \log x \end{cases} \tag{1.5}$$

Last relationship in Eq. 1.5 has a general form of a linear function as presented by $\log y$, and the slope of this linear logarithmic function is the exponential of power law α, and it is known as *Hausdorff–Besicovitch* or *fractal dimension* [3].

1.2.1 Fractal Dimension

This description is adopted from Komulainen [4]. Fractals are characterized by three concepts: self-similarity, response of measure to scale, and the recursive subdivision of space. Fractal dimension can be measured by many different types of methods. Similar to all these methods is that they all rely heavily on the power law when plotted to logarithmic scale, which is the property relating fractal dimension to power laws.

One definition of fractal dimension D is the following equation:

$$D = \log_{10}N/\log_{10}(1/R) \tag{1.6}$$

where N is the number of segments created, when dividing an object, and R is the length of each of segments.

This equation relates to power laws as follows:

$$\log(N) = D \cdot \log(t/R) = \log\left(R^{-D}\right) \tag{1.7}$$

so that

$$N = R^{-D} \tag{1.8}$$

It is simple to obtain a formula for the dimension of any object provided. The procedure is just to determine in how many parts, it is divided up into (N) when we reduce its linear size, or scale it down ($1/R$).

By applying the equation to line, square, and cubicle, we get the following results:

For a line divided in four parts, N is 4, and R is 1/4, so dimension $D = \log(4)/\log(4) = 1$. For a square divided in four parts, N is 4, R is 1/2, and dimension $D = \log(4)/\log(2) = 2 \cdot \log(2) = 2$. And for a cubicle divided in eight parts, N is 8, R is 1/2, and dimension $\log(8)/\log(2) = 3 \cdot \log(2)/\log(2) = 3$.

The following series of pictures (Fig. 1.1) represents iteration of the *Koch curve*. By applying Eq. 1.7 to the Koch curve, as in Table 1.1, it is evident that the dimension is not an integer, but instead between 1 and 2.

For example, if we assume an object has *Hausdorff–Besicovitch* dimension of 1.28, then this object has power-law characteristic that resides in a dimension larger than 1.0 (such as a line), but lower than 2.0 (such as a plane). In other words, the characteristic occupies more than one dimension but less than two dimensions; it occupies a fractional dimension, and it is said to have a *fractal dimension* of 1.28.

Fig. 1.1 The Koch curve

Table 1.1 Statistics of the Koch curve

Iteration number	1	2	3	4	5
Number of segments, N	4	16	64	256	1024
Segment length, R	1/3	1/9	1/27	1/81	1/243
Total length, N, R	1.33333	1.77777	2.37037	3.16049	4.21399
log(N)	0.60206	1.20412	1.80618	2.40824	3.01030
log(1/R)	0.47712	0.95424	1.43136	1.90848	2.38561
Dimension log N/log(1/R)	1.26187	1.26185	1.26185	1.26185	1.26185

We can assume that power law is one of the common footprints of a nonlinear dynamical process, which is a point of self-similarity which is a boundary between order and disorder [5].

The formulas above indicate that N and R are related through a power law.

In general, a power law is a nonlinear relationship, which can be written in the form $N = a(1/R)D$, where D is normally a non-integer constant and a is a numerical constant which in the case of the Koch curve is 1.26. Another way of defining the fractal dimension is box counting. In box counting the fractal is put on a grid, which is made of identical squares having size of side h. Then the amount of nonempty squares, k, is counted. The magnification of the method is equal to $1/h$, and the fractal dimension is defined by equation [5, 6].

$$D = \log_{10}(k)/\log_{10}(1/h) \tag{1.9}$$

In addition, Hausdorff's and Kolmogorov's methods can be used to approximate the fractal dimension. These methods are more accurate, but also harder to use. They are described in [5, 7, 8].

Later on in this book (Chap. 2), we will talk about a system with a self-similar property that statistically shows similar characteristic when examined at the level of individual components of the system, collection of all these components, or the entire system as a whole.

What we basically are trying to define is that the same general characteristic can be seen locally or globally and therefore is independent of the scale at which the observation is made. Such observation allows assuming the hypothesis that the scale-independent nature of a self-similar property is characterized by a power-law distribution. This is because it exhibits structure not merely in response to inputs from outside but also, indeed primarily, in response to its own internal processes [5, 9].

At the other extreme, there are more subtle scaling laws that are used to solve very deep and complicated problems at the frontiers of scientific research. The importance of scaling continues to the present day. However, scaling laws are not merely some particularly simple cases of more general relations. They are of special and exceptional importance. Scaling laws always reveal an important property of the phenomenon under consideration, which is its *self-similarity*. The word "self-similarity" means that a phenomenon reproduces itself on different time and/or space scales [10].

You may be familiar with a simple form of scaling in connection with scale models, such as in Figs. 1.1 and 1.2a, b. (The two figures are the same, except that one has a larger scale than the other does.) The same word shows up in connection with small-scale and large-scale maps.

On the other hand, associating the scaling with similarity, Fig. 1.3 is another good example.

Fig. 1.2 (**a**) Small-scale model train; (**b**) large-scale model train

Fig. 1.3 Scaling and similarity

Scaling laws are a concept in science and engineering. It refers to variables which change drastically depending on the scale (size) being considered. For example, if you tried to build a 50-t mining vehicle using the same engineering assumptions as a 2-t car, you would probably end up with a vehicle that does not even run. The term "scaling laws" often appears when considering the design of a construct that is unusually large or small, so that careful thought is necessary to extend principles of typical-sized constructs to unusually sized constructs.

Some scaling laws are simple. For instance, "for a three-dimensional construct, volume increases with the cube of linear dimensions." This simply means that for every 10 times increase in linear dimensions, the construct's volume increases by a factor of 1000. This is significant for designing machines or structures: if you wanted to double the capacity of a water tower, you would only increase its linear dimensions by a few dozen percent, rather than doubling them. Simple but true.

There are more complex variations of scaling laws where some of the most interesting manifestations of scaling laws are being found in the areas of *micro-technology* and *nanotechnology*, where engineers must both cope with and exploit unusual properties resulting from small scales. In *microfluidics*, some of these unusual properties include laminar flow, surface tension, *electrowetting*, fast thermal relaxation, electrical surface charges, and diffusion. For instance, in fluid chambers with sizes smaller than about half a millimeter, the flow is laminar, meaning that two converging channels cannot mix through turbulence, as on the macroscale, and must instead mix through diffusion.

Other good examples that can be seen using power law and self-similarity are approximations of the Sedov self-similar solution for a strong point explosion in a medium with the power-law density distribution. Petruk [11] reviewed such distribution where he presents power-law density distribution $\rho^o \propto r^{-m}$, and their accuracy are analyzed. In this case he extends the Taylor approximation to cases $m \neq 0$, and two approximations of the solution in the Lagrangian coordinates for spherical, cylindrical, and plane geometry are presented. These approximations may be used for the investigation of the ionization structure of the adiabatic flow, i.e., inside adiabatic supernova remnants.

Application of self-similar solutions for the strong point explosion in the uniform medium of $\widetilde{\rho}^o = \text{constant}$ or in a medium with power-law density distribution is widely used for modeling the adiabatic supernova remnants, solar flares, and processes in active galactic nuclei using relationship of $\widetilde{\rho}^o(\widetilde{r}) = \widetilde{\rho}^o r^{-m}$, where $\widetilde{r}$ is being the distance from the center of explosion. Taylor and Sedov both independently have obtained exact mathematical solution of this problem, by solving the system of hydrodynamic differential equation using the dimensional method [12–14].

There are many other examples of scaling laws.

Example 1.1: Science Using the BEST Data This example is adopted from Sylvan Katz [5], which represents recognition vs. size for world science system. In this example he explores the Matthew effect in science using the BEST data. He shows that in four UK sectoral, six Organization for Economic Cooperation and

Development (OECD) national, a regional and the world science systems the Matthew effect can be described by a power-law relationship between publishing size (papers) and recognition (citations). He demonstrates that the Matthew effect in science (Merton, 1968; Merton, 1988) [15, 16] and other structural features of the science system are self-similar from the level of a sectoral, domestic, and regional science system through to the level of the world science system.

The effect was named the “Matthew effects” in science by Robert Merton in his science paper in 1968 [15], where recognition appears to accumulate with increasing presence in science.

“This classical observation in the sociology of science is based on the general observation that those with a large presence in a community gain more recognition compared to those with little presence or as the old adage says, the rich get richer while the poor get poorer. In other words, as scientists and scientific institutions participate in the science system they gain an accumulative advantage that brings them increasing rewards [5].”

The methodology used by J. Sylvan Katz [5] and conclusion presented by him are clear application of power-law relationship between recognition, c, and publishing size, p, which is given by

$$c = k_a p^{1.27 \pm 0.03}$$

where the intercept, $k_a = 0.15 \pm 0.04$, and the coefficient of determination, r^2, are 0.92.

He used the Matthew effect in science using the BEST data. Figure 1.4 presented below is a scattergram plotted on a log–log scale of publishing size and recognition for 152 communities in the world science system.

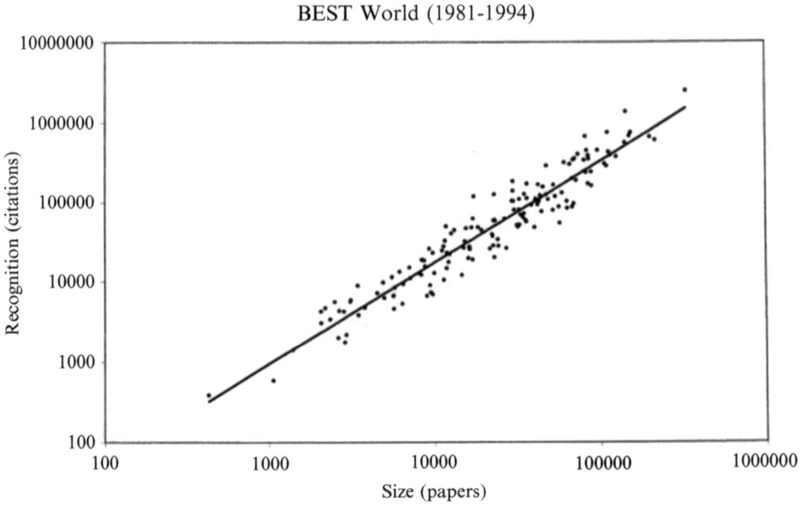

Fig. 1.4 Recognition vs. size for world science system (*Data source*: is from ISI) [5]

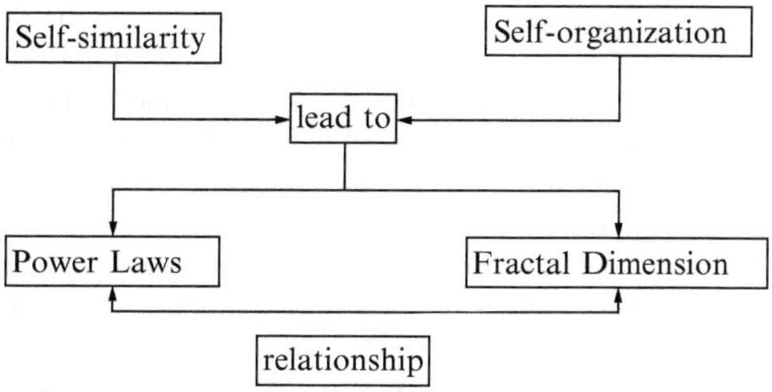

Fig. 1.5 Relationships between self-similarity, self-organization, power laws, and fractal dimension [4]

Size was measured using referred papers published between 1981 and 1992, and recognition was measured by counting citations to those papers between 1981 and 1994 calculated using a 3-year citation window. The straight line on the graph was derived using a regression analysis of a power-law function through the data points. For more details of this example, we encourage the reader to refer to J. Sylvan Katz publication [5].

Note that self-similarity and self-organization to power-scaling laws and to fractal dimension and to show how all these are intertwined together, and it can be depicted in Fig. 1.5 as follows:

Power law is one of the common signatures of a nonlinear dynamical process, i.e., a chaotic process, which is at a point self-organized. With power laws it is possible to express self-similarity of the large and small, i.e., to unite different sizes and lengths. In fractals, for example, there are many more small structures than large ones. Their respective numbers are represented by a power-law distribution. A common power law for all sizes demonstrates the internal self-consistency of the fractal and its unity across all boundaries. The power-law distributions result from a commonality of laws and processes at all scales [4].

The natural world is full of power-law distributions between the large and small: earthquakes, words of the English language, interplanetary debris, and coastlines of continents. For example, power laws define the distribution of catastrophic events in self-organized critical (SOC) systems. If a SOC system shows a power-law distribution, it could be a sign that the system is at the edge of chaos, i.e., going from a stable state to a chaotic state. With power laws, it is useful to predict the phase of this type of systems. A power-law distribution is also a litmus test for self-organization, self-similarity, and fractal geometries [5].

1.3 Scaling Analysis and Modeling

Scaling analysis is fundamental to predicting the behavior of structures and systems when miniaturized. The subject of scaling analysis deals with a systematic method for nondimensionalizing a system of describing equations for transport or reaction process. Scale analysis is a powerful tool used in the mathematical sciences for the simplification of equations with many terms. First the approximate magnitude of individual terms in the equations is determined. Then some negligibly small terms may be ignored. Associated with mathematics, scaling analysis is offering a mathematical model which describes the behavior of a real-life system in terms of mathematical equations.

These equations represent the relations between the relevant properties of the system under consideration. In these models, we meet with variables and parameters. In variables, we discern between *dependent* and *independent*. For example, in mechanical systems one usually is interested in the positions of the different parts as functions of time, so in these systems the positions act as the dependent variables and time as the independent variable. Parameters are properties like masses, prescribed temperatures, currents, voltages, and friction coefficients. Parameters that can be influenced by the observer are referred to as *adjustable*. The other parameters act as constants in the model. For example, in atmospheric models used in weather forecasting one is interested in properties like temperature and humidity (the dependent variables) as functions of position and time (the independent variables).

Important parameters are then the gravity field and the rotational speed of the earth, and these clearly belong to the class of nonadjustable parameters. The solution of a mathematical model is known if we can determine the relations between dependent and independent variables. Since the solution depends on the values of the adjustable parameters, mathematical models are a powerful tool with which to determine which values of the adjustable parameters yield specific required behavior.

Scaling analysis is a tool that identifies dimensionless parameters whose limiting values either very large or very small permit making certain approximations in solving the describing equations. It is a useful tool for developing perturbation expansion solutions to the describing equations.

For example, when Reynolds number is very small, one can develop an analytical solution for the flow around a sphere falling at its terminal velocity in a Newtonian fluid with the constant physical properties, which is the result of familiar Stokes flow solution for creeping flow over a sphere. One can account for the neglected inertia terms in the equations of motion by considering a perturbation expansion solution to the describing equations in terms of the small Reynolds number [17]. The zeroth-order term in this perturbation expansion corresponds to the Stokes solution for creeping flow. Proudman first worked out the first-order term that accounts for some effects of the inertia terms and Pearson perturbation [18] solutions that are well behaved in the limit of the perturbation parameter becoming very small or very large referred to as *regular perturbation expansions* [17]. Perturbation expansions that are not well behaved in the limit of a perturbation parameter becoming very small or very large are referred to as *singular perturbation expansions*. An example of the latter is very high Reynolds number flows. If one tries to solve the equations of motion in the limit of very large Reynolds numbers by attempting a perturbation expansion in the (small) reciprocal Reynolds number, one cannot properly account for the neglected viscous terms. This is a direct consequence of the reduction in the order of the describing equations when one develops the zeroth-order solution in the reciprocal Reynolds number. To solve singular perturbation expansion problems, one needs to use the method of multiple scales, whereby different scales are used in the inner region, the outer region, and the overlap region between them. Scaling analysis is an invaluable tool for determining when perturbation solutions are possible and in determining the proper scales for the various regions.

For the same reason that scaling analysis is useful in determining the scales and expansion parameters in perturbation analyses, it is useful in assessing potential problems that can occur in solving a system of describing equations numerically. That is, when certain dimensionless groups become very small or very large, problems can be encountered in solving the resulting system of describing equations numerically. For example, when the Reynolds number becomes very large, the viscous effects will be confined to a very thin region approximately the solid boundaries. If one uses a coarse mesh or does not employ a numerical routine with a re-meshing capability, the numerical routine will not provide sufficient resolution near the solid boundaries and thereby either will not run or will provide erroneous results. Scaling analysis can be used to identify these *boundary layer regions* so that a proper numerical method can be employed to solve the problem.

Scaling analysis is particularly useful to an educator who is faced with explaining seemingly unrelated topics such as creeping flows, boundary-layer flows, film theory, and penetration theory. Topics such as these often are developed in textbooks in a rather intuitive manner. Scaling analysis provides a systematic way to arrive at these model approximations that eliminate guesswork; that is, scaling analysis provides an invaluable pedagogical tool for teachers. Disparate topics in transport and reaction processes can be presented in a unified and integrated manner. For example, a region of influence in scaling provides a means for presenting a unified approach to boundary-layer theory in fluid dynamics, penetration theory in heat and mass transfer, and the wall region for confined porous media.

1.4 Mathematical Basis for Scaling Analysis

Scaling analysis has its mathematical foundation in specifically the continuous symmetry group of uniform magnifications and contractions, which is known as Lie group theory. The properties of the latter group are useful when considering the operations involved when we change the units on the quantities that appear in dimensional equations [17].

Scaling laws reveal the fundamental property of phenomena, namely, self-similarity—repeating in time and/or space—that substantially simplifies the mathematical modeling of the phenomena themselves. There are many books dealing with analysis of scaling and of the good one written by G. I. Barenblatt. This book begins from a nontraditional exposition of dimensional analysis, physical similarity theory, and general theory of scaling phenomena, using classical examples to demonstrate that the onset of scaling is not until the influence of initial and/or boundary conditions has disappeared but when the system is still far from equilibrium. Numerous examples from a diverse range of fields, including theoretical biology, fracture mechanics, atmospheric and oceanic phenomena, and flame propagation, are presented for which the ideas of scaling, intermediate asymptotic, self-similarity, and renormalization were of decisive value in modeling [18].

For example, when converting the length unit of centimeters to meters, all quantities expressed totally or partially in terms of length, units (heights, widths, velocities, accelerations, densities, etc.) experience either a uniform magnification or contraction. Knowingly then, all heights become smaller when expressed in terms of meters rather than centimeters, whereas all densities become larger.

Scalar analysis might not be very clear in connection between uniform magnifications and contractions in view of the fact that one is not changing units when one nondimensionalizes a system of equations. Nondimensionalizing a quantity will involve dividing the quantity by another quantity or combination of quantities that should have same units.

Lie Group

In mathematics, a Lie group is a group, which is also a differentiable manifold, with the property that the group operations are compatible with the smooth structure. Lie groups are named after *Sophus Lie*, who laid the foundations of the theory of continuous transformation groups.

Lie groups represent the best-developed theory of continuous symmetry of mathematical objects and structures, which makes them indispensable tools for many parts of contemporary mathematics, as well as for modern theoretical physics.

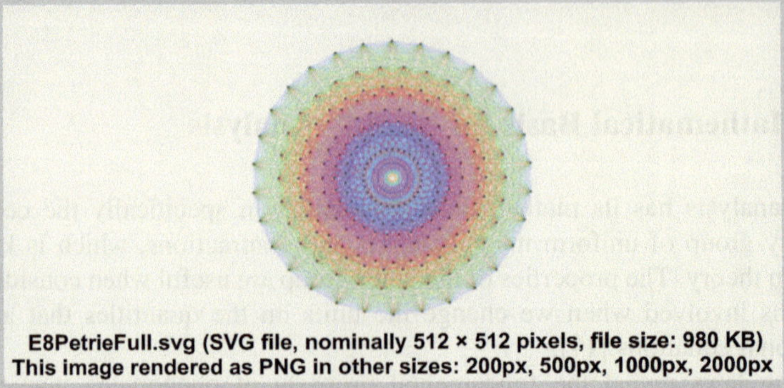

Images for lie group

We are going to learn in the next section that in dimensional analysis, quantities are broadly classified in two kinds that are known as *primary* or *secondary* kind. *Primary quantities* are measured in terms of units of their own kind, for example, a length quantity measured in terms of meters or a force quantity measured in terms of Newton dimension. *Secondary quantities* are measured in terms of the units used for primary quantities, for example, velocity measured in terms of length divided by second squared. Note that any secondary quantity can be converted to a primary quantity merely by measuring it in terms of units of its own kind. Actually, in case of force, the example is considered as both a primary and a secondary quantity. However, the same could be done with a quantity such as velocity. Scaling analysis is equivalent to considering every scaled quantity to be a primary quantity since when we nondimensionalize a quantity, we are dividing it by something that has the same units. Hence, the properties of the group of uniform magnifications and contractions also underlie the operations that we use in scaling analysis [17].

Krantz [17] is suggesting the following steps in order to have a procedure that is involved with scaling analysis of order of one, and these steps are reduced to eight steps.

1. Write the dimensional describing equations and their initial, boundary, and auxiliary conditions appropriate to the transport or reaction process being considered.
2. Define unspecified scale factors for each dependent and independent variable as well as appropriate derivatives appearing explicitly in the describing equations and their initial, boundary, and auxiliary conditions.
3. Define unspecified reference factors for each dependent and independent variable that is not referenced to zero in the initial, boundary, and auxiliary conditions.
4. Form dimensionless variables by introducing the unspecified scale factors and reference factors for the dependent and independent variables and the appropriate derivatives.
5. Introduce these dimensionless variables into the describing equations and their initial, boundary, and auxiliary conditions.
6. Divide through by the dimensional coefficient of one term (preferably one that will be retained) in each of the describing equations and their initial, boundary, and auxiliary conditions.
7. Determine the scale and reference factors by ensuring that the principal terms in the describing equations and initial, boundary, and auxiliary conditions are in order of one. They are bounded between zero and of order one.
8. The preceding steps result in the minimum parametric representation of the problem (i.e., in terms of the minimum number of dimensionless groups); appropriate simplification of the describing equations may now be explored.

1.5 Dimensions, Dimensional Homogeneity, and Independent Dimensions

Returning to the discussion above, recall that if the units of length are changed by a factor of L, and the units of time are changed by a factor of T, then the units of velocity change by a factor of LT^{-1}. We call LT^{-1} the *dimensions* of the velocity; it tells us the factor by which the numerical value of the velocity changes under a change in the units (within the LMT class). Following a convention suggested by Maxwell, we denote the dimensions of a physical quantity ϕ by $[\phi]$; thus, $[v] = LT^{-1}$. A dimensionless quantity would have $[\phi] = 1$, i.e., its numerical value is the same in all systems of units within a given class. What about more complicated quantities such as force? From Newton's second law, $F = ma$, so that $[F] = [m][a] = MLT^{-2}$. Proceeding in this way, we can easily construct the dimensions of any physical quantity; some of the more commonly encountered quantities are included in Table 1.2.

We see that all of the dimensions in the Table are *power-law monomials* of the form (in the *MLTθI*-class)

$$[\phi] = CL^a M^b T^c \tag{1.10}$$

where C and (a, b, c) are constants. In fact, this is a general result which can be proven mathematically; see Sect. 1.4 of G. I. Barenblatt's book [1].

This property is often called *dimensional homogeneity* and is really the key to dimensional analysis. To see why this is useful, consider again the determination of the period of a point pendulum, in a more abstract form. We have for the dimensions $[\omega] = T^{-1}$, $[g] = LT^{-2}$, $[l] = L$, and $[m] = M$.

Table 1.2 Dimensions of some commonly encountered physical quantities in the *LMTθI* class

$[L]$	Length	L
$[M]$	Mass	M
$[T]$	Time	T
$[v]$	Velocity	LT^{-1}
$[a]$	Acceleration	LT^{-2}
$[F]$	Force	MLT^{-2}
$[\rho]$	Mass density	ML^{-3}
$[P]$	Pressure	$ML^{-1}T^{-2}$
$[\alpha]$	Angle	1
$[E]$	Energy	ML^2T^{-2}
$[\theta]$	Temperature	θ
$[S]$	Entropy	$ML^2T^{-2}\theta^{-1}$
$[I]$	Electric current	I
$[Q]$	Electric charge	IT
$[E]$	Electric field	$MLT^{-3}I^{-1}$
$[B]$	Magnetic field	$MLT^{-2}I^{-1}$

If ω is a function of (g, l, m), then its dimensions must be a *power-law monomial of the dimensions* of these quantities. We then have

$$\begin{aligned}[\omega] &= T^{-1} \\ &= [g]^a [l]^b [m]^c \\ &= \left(LT^{-2}\right)^a L^b M^c \\ &= L^{a+b} T^{-2a} M^c\end{aligned}$$

with a, b, and c constants which are determined by comparing the dimensions on both sides of the equation. We see that

$$\begin{cases} a + b = 0 \\ -2a = -1 \\ c = 0 \end{cases}$$

The solution is then $a = 1/2$, $b = -1/2$, and $c = 0$, and we recover Eq. 1.10.

A set of quantities $(a_1, \ldots, a_k)$ is said to have *independent dimensions* if none of these quantities have dimensions, which can be represented as a product of powers of the dimensions of the remaining quantities. As an example, the density ($[\rho] = ML^{-3}$), the velocity ($[v] = LT^{-1}$), and the force ($[F] = MLT^{-2}$) have independent dimensions, so that there is no product of powers of these quantities which is dimensionless. On the other hand, the density, velocity, and pressure ($[p] = ML^{-1}T^{-2}$) are not independent, for we can write $[p] = [\rho][v]^2$, i.e., $p/\rho v^2$ is a dimensionless quantity.

Now suppose we have a relationship between a quantity a which is being determined in some experiment (which we will refer to as the *governed parameter*) and a set of quantities $(a_1, \ldots, a_n)$ which are under experimental control (the governing parameters), which is of the form

$$a = f(a_1, \ldots, a_k, a_{k+1}, \ldots, a_n) \tag{1.11}$$

where $(a_1, \ldots, a_k)$ has independent dimensions. For example, this would mean that the dimensions of the governed parameter a are determined by the dimensions of $(a_1, \ldots, a_k)$, while all of the a_s ' s with $s > k$ can be written as products of powers of the dimensions of $(a_1, \ldots, a_k)$, e.g., $a_{k+1}/a_1^p \ldots a_k^r$ would be dimensionless, with $(p, \ldots, r)$ as an appropriately chosen set of constants. With this set of definitions, it is possible to prove that Eq. 1.11 can be written as

$$a = a_1^p \ldots a_k^r \Phi\left(\frac{a_{k+1}}{a_1^{p_{k+1}} \ldots a_k^{r_{k+1}}}, \ldots, \frac{a_n}{a_1^{p_n} \ldots a_k^{r_n}}\right) \tag{1.12}$$

with Φ some function of dimensionless quantities only. The great simplification is that while the function f in Eq. 1.11 was a function of n variables, the function Φ in Eq. 1.12 is only a function of $n - k$ variables. Equation 1.12 is a mathematical statement of Π theorem, which is the central result of dimensional analysis. The formal proof can again be found in Barenblatt's book [1]. Dimensional analysis cannot supply us with the dimensionless function Φ—we need a real theory for that.

As a simple example of how this works, let us return to the pendulum, but this time we will assume that the mass can be distributed, so that we relax the condition of the mass being concentrated at a point. The governed parameter is the frequency ω; the governing parameters are g, l (which we can interpret as the distance between the pivot point and the center of mass), m, and the moment of inertia about the pivot point, I. Since $[I] = ML^2$, the set of (g, m, l, I) is not independent, then we can choose our independent parameters (g, m, l) as before, with I/ml^2 a dimensionless parameter. In the notation developed above, $n = 4$ and $k = 3$. Therefore, dimensional analysis tells us (Eq. 1.13) that

$$\omega = \sqrt{\frac{g}{l}\Phi\left(\frac{I}{ml^2}\right)} \tag{1.13}$$

with Φ some function, which cannot be determined from dimensional analysis alone, we need a *theory* in order to determine it.

1.6 Basics of Buckingham's π (Pi) Theorem

The *Buckingham π (Pi) theorem* is a key theorem in dimensional analysis, and it is a formalization of *Rayleigh's method of dimensional analysis.* Buckingham in 1941 formulated a theorem, which states that the number of π (Pi) quantities remaining after performing a dimensional analysis is equal to the difference between the number of quantities entering the problem and the maximum number of these that are dimensionally independent. The maximum number of dimensionally independent quantities will always be equal to or less than the number of fundamental dimensions needed to write all dimensional equations.

Another resource that defines this theorem comes from Wikipedia with following description:

> The theorem loosely states that if we have a physically meaningful equation, involving a certain number, n, of physical variables, and these variables are expressible in terms of k independent fundamental physical quantities, then the original expression is equivalent to an equation involving a set of $p = n - k$ dimensionless parameters constructed from the original variables: it is a scheme for nondimensionalization. This provides a method for computing sets of dimensionless parameters from the given variables, even if the form of the equation is still unknown. However, the choice of dimensionless parameters is not unique: Buckingham's theorem only provides a way of generating sets of dimensionless parameters and will not choose the most "physically meaningful."

Other good reference comes from Harald Hanche-Olsen under title of Buckingham's Pi theorem in his mathematical modeling report [19].

As rule of thumb the Buckingham Pi theorem falls into the following steps:

1. The Buckingham Pi theorem is a rule for deciding how many dimensionless numbers (called π's) to expect. The theorem states that the number of independent dimensionless groups is equal to the difference between the number of variables that go to make them up and the number of individual dimensions involved. The weakness of the theorem—from a practical point view—is that it does not depend on the number of dimensions actually used, but rather on the minimum number that might have been used.
2. Firstly, one must decide what variables enter the problem. Occasionally, a dimensional analysis will show that one of the selected variables should not be present, since it involves a dimension not shared by any of the other variables; but if the wrong variables go in, the wrong dimensionless numbers come out, most of the time.
3. One error to avoid in choosing the variables is the inclusion of variables whose influence is already implicitly accounted for. In analyzing the dynamics of a liquid flow, for example, one might argue that the liquid temperature is a significant variable. It is important, however, only in its influence on other properties such as viscosity and should therefore not be included along with them.
4. The Buckingham Pi theorem, if applied to the actual number of dimensions being used, tells only that there must be at least a certain number of dimensionless numbers involved. Unless one resorts to one of the tedious techniques that have been devised for discovering the minimum number of dimensions needed, the theorem gives little assurance that all the dimensionless numbers have been found—an assurance that can very quickly be secured from the step-by-step approach, if assurance is needed.
5. The method of dimensional analysis is based on the obvious fact that in an equation dealing with any system, each term must have the same dimension. For example, if

$$\psi + \eta + \zeta = \varphi$$

is a physical relation, then ψ, η, ζ and φ must have the same dimensions. The above equation can be made dimensionless by dividing by any one of the terms, say φ:

$$(\psi/\varphi) + (\eta/\varphi) + (\zeta/\varphi) = 1$$

These ideas are embodied in the Buckingham Pi theorem, stated below:

(a) Let K equals the number of fundamental dimensions required to describe the physical variables (e.g., mechanics: mass, length, and time; hence $K = 3$).
(b) Let $P_1, P_2, \cdots, P_N$ representing N physical variables in the physical relation $f_1(P_1, P_2, \cdots, P_N) = 0$.

(c) Then, this physical relation may be expressed as a relation of $(N - K)$ dimensionless products (called Π product) and is presented as follows:

$$f_2(\Pi_1, \Pi_2, \cdots, \Pi_{N-K}) = 0$$

where each Π product is a dimensionless product of a set of K physical variables plus one other physical variable.

(d) Let $P_1, P_2, \cdots, P_K$ be the selected set of K physical variables. Then

$$\begin{aligned}
\Pi_1 &= f_3(P_1, P_2, \cdots, P_K, P_{K+1}) \\
\Pi_2 &= f_4(P_1, P_2, \cdots, P_K, P_{K+2}) \\
&\cdots\cdots\cdots\cdots\cdots\cdots\cdots\cdots \\
\Pi_{N-K} &= f_5(P_1, P_2, \cdots, P_K, P_N)
\end{aligned}$$

(e) The choice of the repeating variables $P_1, P_2, \cdots, P_K$ should be such that they include all the K dimensions used in the problem. In addition, the dependent variable should appear in only one of the Π products (Table 1.3).

Summarizing the Buckingham's Pi theorem, we quote Harald Hance [19]. A*ny physically meaningful relation* $\Phi(R_1, \cdots, R_n) = 0$, *with* $R_j \neq 0$, *is equivalent to relation of the form* $\Psi(\pi_1, \cdots, \pi_{n-r}) = 0$ *involving a maximal set of independent dimensionless combinations.*

The important fact to notice is that the new relation involves r fewer variables than the original relation does; this simplifies the theoretical analysis and experimental design alike. Then he goes on to give a precise meaning to the phrase "physically meaningful," which was mentioned above. He does that with assumption of physical quantities $R_1, \cdots, R_n$ where we like to measure them in a *consistent* system of units, such as the SI system, in which the basic units are the meter, kilogram, second, ampere, and kelvin (m, kg, s, A, K) like Table 1.2 and summarizes into simple theory.

Table 1.3 System of units

Sl no	Quantity	Unit	Basic unit	Dimension
1	Force	Newton	kg m/s^2	MLT^2
2	Energy	Joule	kg m^2/s^2	
3	Power	Watt	J/s	
4	Pressure	Pascal	N/m^2	
5	Density	kg/m^3	kg/m^3	ML^{-3}
6	Velocity	M/s	M/s	LT^{-1}
8	Length	M	m	L
9	Viscosity			$ML^{-1}T^{-1}$

Theory

This note is about physical quantities $R_1, \cdots, R_n$. We like to measure them in a in a *consistent* system of units, such as the SI system, in which the basic units are the meter, kilogram, second, ampere, and kelvin (m, kg, s, A, K). As it will turn out, the existence of consistent systems of our system of units is $F_1, \cdots, F_m$, so that we can write

$$R_j = \upsilon(R_j)[R_j] = \rho_j[R_j] \tag{1.14}$$

where $\rho_j = \upsilon(R_j)$ is a number and $[R_j]$ the units of R_j. We can write $[R_j]$ in terms of the fundamental units as a product of powers:

$$[R_j] = \prod_{i=1}^{m} F_i^{a_{ij}} (j = 1, \cdots, n)$$

It is also important for the fundamental units to be *independent* in the sense that

$$\prod_{i=1}^{m} F_i^{ax} = 1 \Rightarrow x_1 = \cdots = x_m = 0 \tag{1.15}$$

We should not be just satisfied with just one system of units: the whole fact of the matter hinges on the fact that our choice of fundamental units is quite arbitrary. Therefore, we might prefer a different system of units, in which the units F_i are replaced by $\hat{F}_i = x_i^{-1} F_i$. Here x_i can be an arbitrary positive number for $i = 1, \cdots, m$. We can also write our quantities in the new system thus: $R_j = \hat{\upsilon}(R_j)[\hat{R}_j] = \hat{\rho}_j[\hat{R}_j]$.

We can compute

$$R_j = \upsilon(R_j) F_1^{a_{1j}} \cdots F_m^{a_{mj}} = \underbrace{\upsilon(R_j) x_1^{a_{1j}} \cdots x_m^{a_{mj}}}_{\hat{\upsilon}(R_j)} \hat{F}_1^{a_{1j}} \cdots \hat{F}_m^{a_{mj}}$$

from which we deduce the relation of Eq. 1.16.

$$\hat{\rho}_j = \rho_j \prod_{i=1}^{m} x_i^{a_{ij}} \tag{1.16}$$

For example, if $F_1 = m$ and $F_s = s$, and R_1 is a velocity, then $[R_1] = ms^{-1} = F_1 F_2^{-1}$ and so $a_{11} = 1$, and $a_{21} = -1$. With $\hat{F}_1 = km$ and $\hat{F}_2 = h$, we find

(continued)

(continued)

$x_1 = 1/1000$ and $x_2 = 1/3600$ and so $\hat{\rho}_1 = (\rho_1)(3.6)$. Hence the example $\rho_1 = 10, \hat{\rho}_1 = 36$ corresponds to the relation $10\ \text{m/s} = 36\ \text{km/h}$.

We define the *dimension matrix* A of $R_1, \cdots, R_n$ by

$$A = \begin{pmatrix} a_{11} & \cdots & a_{1n} \\ \vdots & \ddots & \vdots \\ a_{m1} & \cdots & a_{nm} \end{pmatrix}$$

the *dimension combinations* of the variables R_j. A *combination* of these variables is merely a product of powers $R_1^{\lambda_1} \cdots R_n^{\lambda_n}$. We compute the units of this combination as

$$\left[R_1^{\lambda_1} \cdots R_n^{\lambda_n}\right] = \prod_{i=1}^{m} F_i^{a_{i1}\lambda_1 + \cdots + a_{in}\lambda_n} \tag{1.17}$$

We call the combination *dimensionless* if this unit is 1; thus, we arrive at the important result that this is equivalent to $A\lambda = 0$, where we write $\lambda = (\lambda_1, \cdots, \lambda_n)^T$. *There is, therefore, a one-to-one corresponding between the null space N(A) and the set of dimensionless combination of the variables.*

It may not come as a big surprise that dimensionless combinations have a value independent of the system of units. We simply use Eq. 1.16 and compute

$$\prod_{j=1}^{n} \hat{\rho}_j^{\lambda_j} = \prod_{j=1}^{n} \left(\rho_j \prod_{i=1}^{m} x_i^{a_{ij}}\right)^{\lambda_f} = \left(\prod_{j=1}^{n} \rho_j^{\lambda_f}\right) \prod_{j=1}^{n} \prod_{i=1}^{m} x_i^{a_{if}\lambda_f}$$
$$= \left(\prod_{j=1}^{n} \rho_j^{\lambda_f}\right) \prod_{i=1}^{m} \prod_{j=1}^{n} x_i^{a_{if}\lambda_f} = \prod_{j=1}^{n} \rho_j^{\lambda_f}$$

since $A\lambda = 0$ implies $\prod_{j=1}^{n} x_i^{a_{if}\lambda_f} = 1$.

Moreover, if you pick a basis for $N(A)$ and take the corresponding dimensionless combinations, $\pi_1, \cdots, \pi_{n-r}$ (here r is the rank of A), then any dimensionless combination can be written as a product $\pi_1^{c_1} \cdots \pi_{n-r}^{c_{n-r}}$ where the exponents are uniquely given (they are the coefficients of a member of $N(A)$ in the chosen basis). We shall call this a *maximum set of independent dimensionless combinations*. We now can state *Buckingham's Pi theorem.*

First of all, Φ must also have units, and value:

$$[\Phi] = \prod_{i=1}^{m} F_i^{b_i} \tag{1.18}$$

The value is given by just inserting the values of R_j in the formula for Φ and computing

$$v(\Phi(R_1, \cdots, R_n)) = \Phi(v(R_1), \cdots, v(R_n))$$

Furthermore, when we change to a different set of units, the value of Φ must change according to a law similar to Eq. 1.17 within above-defined *theory*. Thus, we obtain

$$\begin{aligned} \Phi(\hat{v}(R_1), \cdots, \hat{v}(R_n)) &= \hat{v}(\Phi(R_1, \cdots, R_n)) \\ &= x_1^{b_1} \cdots x_m^{b_m} v(\Phi(R_1, \cdots, R_n)) \\ &= x_1^{b_1} \cdots x_m^{b_m} \Phi(v(R_1, \cdots, R_n)) \end{aligned}$$

and therefore

$$\Phi\left(x_1^{a_{11}} \cdots x_m^{a_{m1}} \rho_1, \cdots x_1^{a_{1n}} \cdots x_m^{a_{mn}} \rho_n\right) = x_1^{b_1} \cdots x_m^{b_m} \Phi(\rho_1, \cdots, \rho_n) \tag{1.19}$$

for all real $\rho_1, \cdots, \rho_n$ and positive $x_1, \cdots, x_m$. It is this relation that we shall think of when we say *physically meaningful* in Buckingham's theorem. We shall, however, have to insist on one more feature: since Φ is supposed to combine the quantities R_j, the units of Φ must be the units of some combination of the variable R_j.

Harald Hance Olsen [19] now begins the proof. First, note that, by the final statement of the above paragraph, we may as well replace Φ by $R_1^{c_1} \cdots R_n^{c_n} \Phi(R_1, \cdots, c_1)$ where the coefficients $c_1, \cdots, c_1$ are chosen so that the new function is dimensionless, that is, $b_1 = \cdots = b_m = 0$ in Eq. 1.18.

The dimension matrix A, having the rank r, has r linearly independent columns. We may as well assume these are the first r columns, corresponding to the variables $R_1, \cdots, R_r$. Then $R_1, \cdots, R_r$ are *dimensionally independent* in the sense that their only dimensions are the trivial one: $R_1^{\lambda_1} \cdots R_r^{\lambda_r}$ is dimensionless only if $\lambda_1 = \cdots = \lambda_r = 0$ (this follows immediately from Eq. 1.17). As it was claimed for a natural one-to-one correspondence:

$$(R_1, \cdots, R_n) \leftrightarrow (R_1, \cdots, R_n, \pi_1, \cdots, \pi_{n-r})$$

Clearly, the only possible difficulty here is expressing R_k (where $k > r$) in terms of the quantities on the right-hand side. However, linear algebra tell us that column k of A is a linear combination of the first r columns, and so $[R_k] = \left[R_1^{-c_1} \cdots R_r^{-c_r}\right]$ for a suitable choice $c_1, \cdots, c_r$. Then $R_k = R_1^{-c_1} \cdots R_r^{-c_r}$ is dimensionless, so it can be written as $\pi_1^{d_1} \cdots \pi_{n-r}^{d_{n-r}}$. Therefore, we can write $R_k = \left(R_1^{-c_1} \cdots R_r^{-c_r}\right)\left(\pi_1^{d_1} \cdots \pi_{n-r}^{d_{n-r}}\right)$. Now, using the above one-to-one correspondence, write

$$\Phi(R_1, \cdots, R_n) = (R_1, \cdots, R_n, \pi_1, \cdots, \pi_{n-r}) \tag{1.20}$$

for a suitable function ψ.

In a momentum, we can prove that $\psi(R_1, \cdots, R_n, \pi_1, \cdots, \pi_{n-r})$ does not depend on $R_1, \cdots, R_n$. Thus we may write

$$\psi(R_1, \cdots, R_n, \pi_1, \cdots . \pi_{n-r}) = \psi(\pi_1, \cdots, \pi_{n-r})$$

Moreover, the proof of *Buckingham's Pi theorem* will be complete.

To prove the independence of $R_1, \cdots, R_r$, replace each in Eq. 1.20 by its value ρ_j and substitute this in both sides of Eq. 1.13 and remember that $b_i = 0$.

$$\begin{aligned}&\psi\left(x_1^{a_{11}} \cdots x_m^{a_{m1}} \rho_1, \cdots, x_1^{a_{mr}} \rho_r, \pi_1, \cdots, \pi_{n-r}\right)\\&= \psi(\rho_1, \cdots, \rho_r, \pi_1, \cdots, \pi_{n-r})\end{aligned}$$

Now, we claim that, given positive numbers $\rho_1, \cdots, \rho_r$, we can pick positive numbers $x_i, \cdots, x_m$ so that the numbers $x_1^{a_{11}} \cdots x_m^{a_{m1}} \rho_j$ for $j = 1, \cdots, r$ on the left-hand side of the equation can be any given positive numbers. To be specific, we can make them all equal to 1. That is, we can solve the equations

$$\prod_{i=1}^{m} x_i^{a_{ij}} = \frac{1}{\rho_j} j = 1, \cdots, r$$

with respect to x_i. In fact, if we write $x_i = \exp(\xi_i)$, the above equation is equivalent to

$$\sum_{i=1}^{m} a_{ij} \xi = -\ln \rho_j \quad j = 1, \cdots, r$$

This equation is solvable because the left $m \times r$ sub-matrix of A has rank r, and therefore its rows span $\mathbb{R}^r$. This proves the claim above and therefore the theorem. Note that in most books, $m = n$ and $r = k$, which will be used in our following examples to demonstrate *Buckingham's Pi theorem* applications.

1.6.1 Some Examples of Buckingham's π (Pi) Theorem

Example 1.2: Speed This example is elementary, but demonstrates the general procedure. Suppose a car is driving at 100 km/h; how long does it take it to go 200 km? This simple example is borrowed from Wikipedia.

This question has two fundamental physical units: time t and length l and three-dimensional variables, as follows:

1. D for distance
2. T for time taken
3. V for velocity

Thus, $p = n - k$ is going to be $3 - 2 = 1$ dimensionless quantity. The units of the dimensional quantities are

$$D \propto l, \quad T \propto t, \quad V \propto \frac{l}{t}$$

The dimensional matrix then is

$$M = \begin{bmatrix} 1 & 0 & 1 \\ 0 & 1 & -1 \end{bmatrix}$$

The rows correspond to the dimensions l and t and the columns to the dimensional D, T, and V. For instance the third column, (1,−1), states that the V (velocity) variable has units of $l^1 t^{-1} = l/t$.

For a dimensional constant $\pi = D^{a_1} T^{a_2} V^{a_3}$, we are looking for a vector $a = [a_1, a_2, a_3]$ such that the matrix product of M on a yields the zero vector [0,0]. In linear algebra, this vector is known as the *kernel* of the dimensional matrix, and it spans the null space of the dimensional matrix, which in this particular case is one-dimensional. The dimensional matrix as written above is in reduced row echelon form, so one can read off that a kernel vector may be written (to within multiplicative constant) by

$$a = \begin{bmatrix} -1 \\ 1 \\ 1 \end{bmatrix}$$

If the dimensional matrix were not already reduced, one could perform *Gauss–Jordan elimination* on the dimensional matrix in order to more easily determine the kernel. It follows that the dimensionless constant be written:

$$\begin{aligned} \pi &= D^{-1} T^1 V^1 \\ &= \frac{TV}{D} \end{aligned}$$

Alternatively, in dimensional terms

$$\pi \sim (l)^{-1} (t)^1 (l/t)^1 \sim 1$$

Since the kernel is only defined to within a multiplicative constant, if the above dimensionless constant is raised to any arbitrary power, it will yield another equivalent dimensionless constant. Dimensional analysis has thus provided a general equation relating the three physical variables:

$$f(\pi) = 0$$

This may be written as

$$T = \frac{CD}{V}$$

where C is one of a set of constants, such that $C = f^{-1}(0)$. The actual relationship between the three variables is simply $D = VT$ so that the actual dimensionless equation $[f(\pi) = 0]$ is written:

$$f(\pi) = \pi - 1 = \left(\frac{VT}{D}\right) - 1 = 0$$

In other words, there is only one value of C, and it is unity. The fact that there is only a single value of C and that it is equal to unity is a level of detail not provided by the technique of dimensional analysis.

Example 1.3: The Simple Pendulum Motion This example, mathematically with more details, has been introduced in Appendix A of the book. Now we take the approach that Wikipedia is showing (Fig. 1.6).

We wish to determine the period T of small oscillations in a simple pendulum. It will be assumed that it is a function of the length L, the mass M, and the acceleration due to gravity on the surface of the Earth g, which has units of length divided by time squared. The model is of the form

Fig. 1.6 Simple pendulum

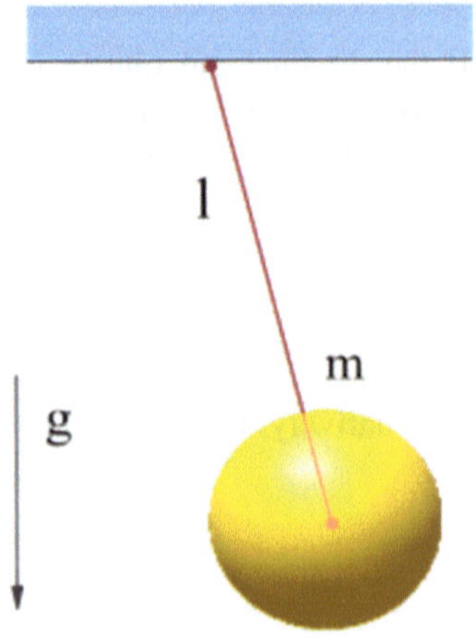

$$f(T, M, L, g) = 0$$

(Note that it is written as a relation, not as a function: T is not here written as a function of M, L, and g.)

There are three fundamental physical units in this equation, time t, mass m, and length l, and four dimensional variables, T, M, L, and g. Thus, we need only $4 - 3 = 1$ dimensionless parameter, denoted p, and the model can be re-expressed as

$$f(\pi) = 0$$

where π is given by

$$\pi = T^{a_1} M^{a_2} L^{a_3} g^{a_4}$$

for some values of $a_1, \ldots a_4$.

The units of the dimensional quantities are

$$T = t, \quad M = m, \quad L = l, \quad g = \frac{l}{t^2}$$

The dimensional matrix is

$$M = \begin{bmatrix} 1 & 0 & 0 & -2 \\ 0 & 1 & 0 & 0 \\ 0 & 0 & 1 & 1 \end{bmatrix}$$

(The rows correspond to the dimensions t, m, and l and the columns to the dimensional variables T, M, L, and g. For instance, the fourth column, (−2, 0, 1), states that the g variable has units of $t^{-2}m^0l^1$).

We are looking for a kernel vector a = [a1, a2, a3, a4] such that the matrix product of M on a yields the zero vector [0,0,0]. The dimensional matrix as written above is in reduced row echelon form, so one can read off that a kernel vector may be written (to within a multiplicative constant) by

$$a = \begin{bmatrix} 2 \\ 0 \\ -1 \\ 1 \end{bmatrix}$$

Where it is not already reduced, one could perform *Gauss–Jordan elimination* on the dimensional matrix in order to more easily determine the kernel. It follows that the dimensionless constant may be written:

$$\begin{aligned} \pi &= T^2 M^0 L^{-1} g^1 \\ &= gT^2/L \end{aligned}$$

In dimensional terms

$$\pi = (t)^2(m)^0(l)^{-1}\left(l/t^2\right)^1 = 1$$

This is dimensionless. Since the kernel is only defined to within a multiplicative constant, if the above dimensionless constant is raised to any arbitrary power, it will yield another equivalent dimensionless constant.

This example is easy because three of the dimensional quantities are fundamental units, so the last (g) is a combination of the previous. Note that if a_2 was nonzero, there would be no way to cancel the M value; therefore, a_2 must be zero. Dimensional analysis has allowed us to conclude that the period of the pendulum is not a function of its mass. (In the three-dimensional space of powers of mass, time, and distance, we can say that the vector for mass is linearly independent from the vectors for the three other variables. Up to a scaling factor, $\vec{g} - 2\vec{T} - \vec{L}$ is the only nontrivial way to construct a vector of a dimensionless parameter.)

The model can now be expressed as

$$f\left(gT^2/L\right) = 0$$

Assuming the zeroes of f are discrete, we can say $gT^2/L = C_n$, where C_n is the nth zero. If there is only one zero, then $gT^2/L = C$. It requires more insight that is physical or an experiment to show that there is indeed only one zero and that the constant is in fact given by $C = 4\pi^2$.

For large oscillations of a pendulum, the analysis is complicated by an additional dimensionless parameter, the maximum swing angle. The above analysis is a good approximation as the *angle approaches zero.*

Example 1.4: The Atomic Bomb Explosion This problem is presented with totally different approach using dimensional analysis method, yet here we use the solution from Wikipedia using the Pi theorem.

Sir Geoffrey I. Taylor used dimensional analysis to estimate the energy released in an atomic bomb explosion (Taylor, 1950a and b) [20, 21]. The first atomic bomb was detonated near Alamogordo, New Mexico, on July 16, 1945. In 1947, movies of the explosion were declassified, allowing Sir Geoffrey to complete the analysis and estimate the energy released in the explosion, even though the energy release was still classified. The actual energy released was later declassified, and its value was remarkably close to Taylor's estimate.

Taylor supposed that the process was adequately described by five physical quantities: the time t since the detonation, the energy E that is released at a single point in space at detonation, the radius R of the shock wave at time T, the atmospheric pressure p, and the ambient density ρ.

There are only three fundamental physical units in this equation: mass, time, and length. Thus we need only $5 - 3 = 2$ dimensionless parameters. The dimensional matrix for rows corresponding to l, t, and m and columns corresponding to R, T, p, ρ, and E is

$$M = \begin{bmatrix} 1 & 0 & -1 & -3 & 2 \\ 0 & 1 & -2 & 0 & -2 \\ 0 & 0 & 1 & 1 & 1 \end{bmatrix}$$

The null space is two-dimensional. Two kernel vectors, which span this space, are

$$A_0 = [5, -2, 0, 1, -1] \quad \text{and} \quad A_1 = [0, 6, 5, -3, -3]$$

which yield the dimensionless constants

$$\pi_0 = R\left(\frac{\rho}{Et^2}\right) \quad \text{and} \quad \pi_1 = p\left(\frac{t^6}{E^2\rho^3}\right)^{1/5}$$

Note that there are an infinite number of dimensionless constant pairs that could be defined by taking linear combinations of the above two particular kernel vectors. The process can now be described by an equation of the form

$$f(\pi_0, \pi_1) = 0$$

Assuming that inverting the equation yields a single possible value for R, this may be written:

$$R = \left(\frac{Et^2}{\rho}\right)^{1/5} g(\pi_1)$$

where $g(\pi_1)$ is some function of π_1. The energy in the explosion is expected to be huge, so that for times of the order of a second after the explosion, we can estimate π_1 to be approximately zero, and experiments using light explosives can be conducted to determine that $g(0)$ is on the order of unity so that

$$R \approx \left(\frac{Et^2}{\rho}\right)^{1/5}$$

This is Taylor's equation, which, once he knew the radius of the explosion as a function of the time, allowed him to calculate the energy of the explosion.

Example 1.5: Motion of a Projectile This problem can be done utilizing the motion of projectile equation that can be found in any elementary college physics book for elevation condition as shown below:

Table 1.4 Dimensional equations for projectile problem

Quantity	Dimensions
Vertical displacement, y	$[L]$
Acceleration due to gravity, g	$[LT^{-2}]$
Time, t	$[T]$
Initial velocity, v_0	$[LT^{-1}]$

$$y = -\frac{1}{2}gt^2 + v_0 t \sin\theta$$

where t is the time elapsed after launching the projectile with an initial velocity, v_0, inclined at an angle θ to the horizontal and g is considered as gravitation force.

In considering this problem by the dimensional analysis method, it would first be observed that the main dependent variable y will be some function ψ_1 of v_0, t, and g, i.e.,

$$y = \psi_1(v_0, t, g)$$

Since this is a kinematics problem, two fundamental dimensions (L, T) are involved.

Table 1.4 gives the fundamental dimensions for all required quantities entering this problem. The maximum number of dimensionally independent quantities should, therefore, be two (g and t).

The quantities g and t are dimensionally independent (may not be combined by raising to powers and multiplying together to form a nondimensional group). After performing the dimensional analysis, the following result would be obtained:

$$\frac{y}{gt^2} = \psi_2(g, t, v_0/gt)$$

The left side of this equation is nondimensional, and by the principle of dimensional homogeneity, all terms of ψ_2 must also be nondimensional. The quantities g and t, therefore, cannot appear in ψ_2 except when combined with other quantities to form a nondimensional group. Thus, we get the following form of above equation:

$$\frac{y}{gt^2} = \psi_2(v_0/gt)$$

Applying Buckingham's Pi theorem to this case, we will see that $n = 4$ and $k = 2$, so π (Pi) quantities are calculated as follows:

$$\pi(\text{Pi})\ \text{quantities} = \text{Total quantities} - \text{Dimensionally independent quantities}$$
$$\pi(\text{Pi})\ \text{quantities} = n - k = 4 - 2 = 2$$

The dimensionally independent quantities can usually be analyzed and selected in more than one way. One can often find more than several correct answers to

Table 1.5 Possible dimensional analysis for projectile problem

Dimensional independent quantities	Resulting equation
g, t	$y/gt^2 = \psi[v_0/gt]$
g, v_0	$(yg)/v_0^2 = \psi[gt/v_0]$
t, v_0	$y/v_0 t = \psi[gt/v_0]$

dimensional analysis. One of these answers may prove to be more convenient for a given purpose than the others. In the projectile problem, all of the pairs of dimensionally independent quantities listed in Table 1.5 could have been used, and the dimensionless equations listed in each pair of dimensionally independent quantities would have been obtained.

Exponents a and b required to make $yg^a t^b$ nondimensionally may be found by writing simultaneous equations as follows (although these exponents may usually be written by inspection):

$$yg^a t^b = \underbrace{\left[L\left(LT^{-2}\right)^a T^b\right]}_{\text{I}} = \underbrace{\left[L^0 T^0\right]}_{\text{II}}$$

Equating exponents of L in I and II yields

$$1 + a = 0$$

while equating the exponents for T in I and II gives

$$-2a + b = 0$$

When these equations are solved simultaneously, a and b are found to be -1 and -2, respectively, in agreement with the left-hand side of equation of motion for projectile.

The result of a dimensional analysis is sometimes written in a symbolic form in terms of nondimensional Pi quantities as follows:

$$\pi_1 = \psi(\pi_2, \pi_3, \text{etc.})$$

where π_1 is a nondimensional group involving the main dependent variable (y in the projectile problem) and the other Pi values represent the remaining nondimensional quantities entering the problem.

Example 1.6: Pipe Flow We consider the problem of determining the pressure drop of a fluid flowing through a pipe. If the pipe is long compared to its diameter, we shall assume that the pressure drop is proportional to the length of the pipe and all other factors being equal. Thus, we really look for the (average) pressure gradient ∇P and presume the length of the pipe to be irrelevant.

Variables that are relevant clearly include other properties of the pipe: its diameter, D, and its roughness, e. To a first approximation, we just let e be the average size of the unevennesses of the inside surface of the pipe; thus it is a length.

Also relevant are fluid properties. We shall use the kinematic viscosity $v = \mu/\rho$ together with the density ρ. In a Newtonian fluid in shear motion, the shear tension (a force per unit area) is proportional to a velocity gradient, and the dynamic viscosity μ is the required constant of proportionality: thus, the units of μ are Nm $^{-2}$/s^{-1} = kg m^{-1} s^{-1}, and therefore the units of ν are m^{2}s^{-1}.

Finally, the average fluid velocity v is most definitely needed.

The dimension matrix can be written as follows:

	∇P	υ	D	e	ν	ρ
M	−2	1	1	1	2 − 3	
kg	1	0	0	0	0	1
s	−2	−1	0	0	−1	0

We can find the null space of this and hence use it to find the dimensionless combinations. However, it is in fact easier to find dimensionless combinations by inspection. It is easy to see that the matrix has rank $k = 3$, so with $n = 6$ variables, we must find $6 - 3 = 3$ independent dimensionless combinations. There are, of course, an infinite number of possibilities, since the choice corresponds to choosing a basis for the null space of A. In this case, we may be guided by common practice, however, and pick dimensionless quantities as follows:

$$\text{Reynold's number} \quad \text{Re} = \frac{\upsilon D}{\nu}$$

$$\text{Relative roughness} \quad \varepsilon = \frac{e}{D}$$

$$\text{(no name relationship)} \quad \frac{\nabla P \cdot D}{\rho \upsilon^2}$$

Since we expect the ∇P to be a function of the other variables, we should have a relationship between the above quantities, which has a unique solution for the only variable containing ∇P:

$$\frac{\nabla P \cdot D}{\rho \upsilon^2} = f(\text{Re}, \varepsilon)$$

which we write as

$$\nabla P = \frac{2\rho \upsilon^2}{D} f(\text{Re}, \varepsilon)$$

The extra factor 2 is there because then f is known as Fanning's friction factor. Presumably, Fanning used the radius of the pipe rather than the diameter as the basis for his analysis. (A copy and description of the *Moody* diagram should be included here). The Moody chart or Moody diagram is a graph in nondimensional form that relates the Darcy friction factor, Reynolds number, and relative roughness

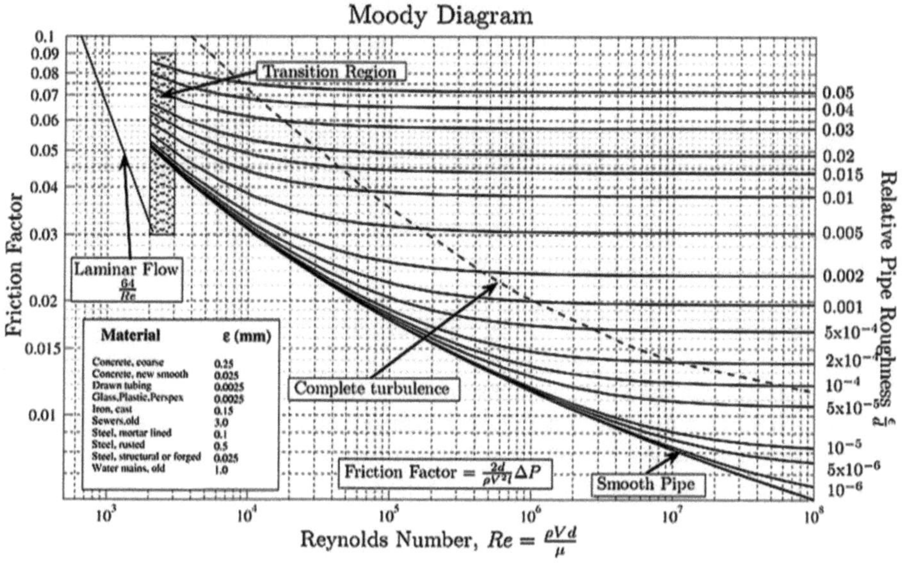

Fig. 1.7 Moody diagram showing the Darcy friction factor plotted against Reynolds number for various roughnesses

for fully developed flow in a circular pipe. It can be used for working out pressure drop or flow rate down such a pipe. A depiction of such diagram is presented below (Fig. 1.7).

As one final remark, we did not really need to write down the dimension matrix. It is quite clear that the three dimensionless quantities we found are independent, since each of them contains at least one variable, which is not present in the two others. Since there were only three fundamental units involved, the dimension matrix could not possibly have rank greater than three, and therefore there could not exist more than three independent dimensionless combinations. Still, the dimension matrix provides a convenient way to summarize the dimensions and to reduce everything to a problem in linear algebra.

Example 1.7: Water Waves We consider surface waves in water. These waves can be conveniently characterized by a wave number $k = 2\pi/\lambda$ (where λ is the wavelength) and an angular frequency ω.We seek a *dispersion relation* expressing ω as a function of ω. Presumably, the depth d plays a role, as well as the wave height h, the acceleration of gravity g, and the fluid properties: the density ρ and (for very small waves) the surface tension τ. We shall assume that the viscosity is negligible.

The dimensions of all these variables can be summarized as follows:

Variable	ω	k	h	d	ρ	τ	g
Units	s^{-1}	m^{-1}	m	m	kg m^{-3}	$Nm^{-1} = kg\ s^{-2}$	ms^{-2}

With no less than seven variables, and three fundamental units, we expect to find four independent dimensionless combinations. One reasonable choice is (Bo is the *Bond number*)

$$hk \quad dk \quad \frac{\omega^2}{gk} \quad \mathrm{Bo} = \frac{\rho g}{\tau k^2}$$

A relationship between all these, solved for the one combination that involves ω, then leads to a relationship of the form

$$\omega^2 = gk\Psi(hk, dk, \mathrm{Bo})$$

We see that, for example, when waves are long, Bo is large, so we may ignore the influence of surface tension. If the water is deep compared to the wave length, then $dk \approx \infty$, while if the wave height is small compared to wavelength, $hk \approx 0$. When all of these approximations hold, then, we expect Ψ to be roughly constant, so ω^2 is proportional to gk. In fact, we find $\omega^2 = gk$ in the limit, but this requires analysis that is more detailed.

For very short waves (ripples) in deep water, it seems reasonable to assume that only surface tension is responsible for the wave motion, so that g does not enter the problem. You could do a new dimensional analysis under this assumption, but it is easier to see directly that Ψ must be a linear function of its last argument for g to cancel out. If we still assume $dk \gg 1$ and $hk \ll 1$, we end up with a relationship of the form

$$\omega^2 = \frac{\tau k^3}{\rho}$$

except the right-hand side should be multiplied by a dimensionless constant. Again, this constant turns out to be 1.

We can finish this section with one final example that is borrowed from Dym [22] book.

Example 1.8: Making Peanut Butter We want to design a mixer machine that will make large quantities of peanut butter, since moving a knife through a jar of peanut butter requires some sort of force than string a glass of what we stick with the recipe that Dym [22] recommends in his book. It turns out, as one may expect, that the forces depend in large part on properties of the peanut butter, but on which properties, and how? He answers these questions by expressing a performance series of operation one may have to do in order to make the peanut butter of his desired recipe. Therefore, the five quantities that we will take under consideration and are derived quantities for this initial investigation into the mixing properties of peanut butter are as follows:

1. Drag force on mixer blades F_{D}
2. The knife blade width, d

3. Speed of blades move and turn, V
4. Peanut viscosity, μ
5. Peanut butter's mass density, ρ

The fundamental physical quantities we would apply are mass, length, and time for this problem, and we donate them as M, L, and T, respectively. The derived variables are expressed in terms of the fundamental quantities in the table below:

Derived quantities	Dimensions
Speed (V)	L/T
Blade width (d)	L
Density (ρ)	$M/(L)^3$
Viscosity (μ)	$M/(L \times T)$
Drag force (F_D)	$(M \times L)/(T)^2$

The five derived peanut butter to model the peanut butter stirring experiments

The drag force, the force that is required to pull the blade through the butter, is directly proportional both to the speed with which it moves and the area of the blade and inversely proportional to a length that characterizes the spatial rate of change of the speed; thus, we have

$$F_D \propto \frac{VA}{L}$$

$$F_D \propto \mu \frac{VA}{L}$$

If we apply the principle of dimensional homogeneous to above equation, it follows that

$$[\mu] = \left[\frac{F_D}{A} \times \frac{L}{V}\right]$$

It is required to know how F_D and V related, and yet they are also functions of the other variables, d, ρ, and μ, and, that is,

$$F_D = F_D(V, d, \rho, \mu)$$

Now using the Pi theorem we show how this problem can be "reduced" to considering two dimensionless groups that are related by a single curve.

There should be two dimensionless groups correlating the five variables of the problem listed in the above table. To apply the Pi theorem to this mixer we, choose the blade speed V, its width d, and the butter density ρ as the fundamental variables ($k = 3$), which we then permute with two remaining variables—the viscosity μ and the drag force F_D—to get two dimensionless groups:

$$\Pi_1 = V^{a_1} d^{b_1} \rho^{c_1} \mu$$
$$\Pi_2 = V^{a_2} d^{b_2} \rho^{c_2} F_D$$

Expressed in terms of primary dimensions, these groups are

$$\Pi_1 = \left(\tfrac{L}{T}\right)^{a_1} L^{b_1} \left(\tfrac{M}{L^3}\right)^{c_1} \left(\frac{M}{LT}\right)$$
$$\Pi_2 = \left(\tfrac{L}{T}\right)^{a_2} L^{b_2} \left(\tfrac{M}{L^3}\right)^{c_2} \left(\frac{ML}{T^2}\right)$$

Now in order for Π_1 and Π_2 to be dimensionless, the exponents for each of the three primary dimensions must vanish. Thus, for Π_1

$$\begin{aligned} L:&\ \ a_1 + b_1 - 3c_1 - 1 = 0 \\ T:&\ \ -a_1 - 1 = 0 \\ L:&\ \ c_1 - 1 = 0 \end{aligned}$$

and for Π_2

$$\begin{aligned} L:&\ \ a_2 + b_2 - 3c_2 + 1 = 0 \\ T:&\ \ -a_2 - 2 = 0 \\ L:&\ \ c_2 + 1 = 0 \end{aligned}$$

Solving the above set of equation for the pairs of subscripts yields

$$a_1 = b_1 = -1 \quad c_1 = -1$$
$$a_2 = b_2 = -2 \quad c_2 = -1$$

Then the two dimensionless groups for the peanut butter mixer are

$$\Pi_1 = \left(\frac{\mu}{\rho V d}\right)$$
$$\Pi_2 = \left(\frac{F_D}{\rho V^2 d^2}\right)$$

Thus, two dimensionless groups should guide experiments with prototype peanut butter mixers. One clearly involves the viscosity of the peanut butter, while the other relates the drag force on the blade to the blade's dimensions and speed, as well as to the density of peanut butter.

1.7 Oscillations of a Star

A star undergoes some mode of oscillation. How does the frequency ω of oscillation depend upon the properties of the star? The first step is the identification of the physically relevant variables. Certainly, the density ρ and the radius R are important; we will also need the gravitational constant G, which appears in Newton's law of universal gravitation. We could add the mass m to the list, but if we assume that the density is constant as a first approximation, then $m = \rho\left(4\pi R^3/3\right)$, and the mass is redundant. Therefore, ω is the governed parameter, with dimensions $[\omega] = T^{-1}$, and (ρ, R, G) are the governing parameters, with dimensions $[\rho] = ML^{-3}$, $[R] = L$, and $[G] = M^{-1}L^3T^{-2}$ (check the last one). You can easily check that (ρ, R, G) have independent dimensions; therefore, $n = 3$and $k = 3$, so the function Φ is simply a constant in this case. Next, determine the exponents

$$\begin{aligned}[\omega] &= T^{-1} \\ &= [\rho]^a[R]^b[G]^c \\ &= M^{a-c}L^{-3a+b+3c}T^{-2c}\end{aligned}$$

Equating exponents on both sides, we have

$$\begin{cases} a - c = 0 \\ -3a + b + 3c = 0 \\ -2c = -1 \end{cases}$$

Solving, we find $a = c = 1/2$and $b = 0$, so that

$$\omega = C\sqrt{G\rho}$$

with C being a constant. We see that the frequency of oscillation is proportional to the square root of the density and independent of the radius. Once again, the determination of C requires a real theory of stellar oscillation, but the interesting dependence upon the physical parameters has been obtained from dimensional considerations alone.

1.8 Gravity Waves on Water

Next consider waves on the surface of water, which are simple *gravity waves*. The effect of surface tension can be neglected (this is valid for long waves, while for short ripples the surface tension is dominant and gives rise to *capillary waves*; gravity and surface tension are equally important at a wavelength of 5 cm). How does the frequency ω depend upon the wave number k (recall that $k = 2\pi/\lambda$, where λ is the wavelength) of the wave? The relationship $\omega = \omega(k)$ is known as the *dispersion relation* for the wave. The relevant variables would appear to be (ρ, g, k), which have dimensions $[\rho] = ML^{-3}$, $[g] = LT^{-2}$ and $[k] = L^{-1}$; these quantities have independent dimensions, so $n = 3$ and $k = 3$. Now we can determine the exponents

$$\begin{aligned}[\omega] &= T^{-1} \\ &= [\rho]^a [g]^b [k]^c \\ &= M^a L^{-3a+b-c} T^{-2b}\end{aligned}$$

so that

$$\begin{cases} a = 0 \\ -3a + b - c = 0 \\ -2b = -1 \end{cases}$$

with the solution $a = 0$ and $b = c = 1/2$. Therefore,

$$\omega = C\sqrt{gk}$$

with C another undetermined constant. We see that the frequency of water waves is proportional to the square root of the wave number, in contrast to sound or light waves, for which the frequency is proportional to the wave number. This has the interesting consequence that the group velocity of these waves is $v_g = \partial\omega/\partial k = (C/2)\sqrt{g/k}$, while the phase velocity is $v_\phi = \omega/k = C\sqrt{g/k}$, so that $v_g = v_\phi/2$. Recall that the group velocity describes the large-scale "lumps" which would occur when we superimpose two waves, while the phase velocity describes the short-scale "wavelets" inside the lumps. For water waves, these wavelets travel twice as fast as the lumps.

You might worry about the effects of surface tension σ on the dispersion relationship. We can include these in our dimensional analysis by recalling that the surface tension is the energy per unit area of the surface of the water, so it has dimensions $[\sigma] = MT^{-2}$. The dimensions of the surface tension are not independent of the dimensions of (ρ, g, k); in fact, it is easy to show that $[\sigma] = [\rho][g][k]^{-2}$, so that $\sigma k^2/\rho g$ is dimensionless. Then using the same arguments as before, we have

$$\omega = \sqrt{gk}\Phi\left(\frac{\sigma k^2}{\rho g}\right)$$

with Φ some undetermined function. A calculation of the dispersion relation for gravity waves starting from the fundamental equations of fluid mechanics [23] gives

$$\omega = \sqrt{gk}\sqrt{1 + \sigma k^2/\rho g}$$

so that our function $\Phi(x)$ is

$$\Phi(x) = \sqrt{1 + x}$$

Dimensional analysis enabled us to deduce the correct form of the solution, i.e., the possible combinations of the variables. Of course, only a complete theory could provide us with the function $\Phi(x)$.

What have we gained? We originally started with ω being a function of the four variables (ρ, g, k, σ); what dimensional analysis tells us is that it is really only a function of the combination $\sigma k^2/\rho g$, even though we do not know the function. Notice that this is an important fact if you are trying to measure the dependence of ω on the physical parameters (ρ, g, k, σ). If you needed to make (say) ten separate measurements on each variable while holding the others fixed, then without dimensional analysis you would naively need to make 10^4 separate measurements. Dimensional analysis tells you that you only really need to measure the combinations gk and $\sigma k^2/\rho g$, so only need to make 10^2 measurements to characterize ω. Dimensional analysis can be a labor-saving device!

1.9 Dimensional Analysis Correlation for Cooking a Turkey

In this example, we are trying to implement the use of scaling analysis for dimensional analysis in heat transfer to develop a correlation for determining the cooking time of a turkey. Let t_c denote the cooking time for the turkey. Now we have to determine dependency variable of t_c. First think that what comes to mind is the size of the turkey, which is a factor for cooking of the turkey. Let us assume l represents the length of the turkey and all the turkeys are geometrically similar and use l to reflect some characteristic dimension of the uncooked turkey meat. Another dependent variable factor that we need to consider is the difference between the temperature of the raw meat and the oven ΔT_m. (*Note* that from experience one should know that it takes longer to cook a frozen turkey than the one that is initially at room temperature.) Because the turkey needs to reach a certain interior temperature before it is consider fully cooked, the difference ΔT_c between the temperature of the cooked meat and the oven is a variable determining the cooking time. Finally, we know that different foods require different cooking times independent of size; it takes only 10 min or so to bake a pan of cookies, whereas a roast beef or turkey requires several hours. A measure of the factor representing the differences between foods is the *coefficient of head conduction* for a particular uncooked food. Let k denote the coefficient of heat conduction for a turkey. Thus, we have the following model formulation for the cooking time:

$$t_c = f(\Delta T_m, \Delta T_c, k, l)$$

Under the above assumption, the process of cooking a turkey using dimensional analysis method seems a fairly straightforward and simple problem. The temperature variables ΔT_m and ΔT_c measure the energy per volume and therefore have the dimension ML^2T^{-2}/L^3 or simply $ML^{-1}T^{-2}$. Now, what we need to analyze is the heat conduction variable k dimension. Thermal conductivity k is defined as the amount of energy crossing one unit cross-sectional area per second divided by the gradient perpendicular to the area. That is,

$$k = \frac{\text{energy}/(\text{area} \times \text{time})}{\text{temperature}/\text{length}}$$

Accordingly, the dimension of k is $(ML^2T^{-2})(L^{-2}T^{-1})/(ML^{-1}T^{-2})(L^{-1})$ or simply L^2T^{-1}. Our analysis gives the following table:

Variable	ΔT_m	ΔT_c	k	l	t_c
Dimension	$ML^{-1}T^{-2}$	$ML^{-1}T^{-2}$	L^2T^{-1}	L	T

Any product of the variables must be of the form of the following type:

$$\Delta T_m^a \Delta T_c^b k^c l^d t_c^e \tag{1}$$

and hence have the dimension of the form

$$(ML^{-1}T^{-2})^a (ML^{-1}T^{-2})^b (L^{-2}T^{-1})^c (L)^d (T)^e$$

Therefore, a product of the form of Eq. 1 above is dimensionless if and only if the exponents satisfy the following conditions:

$$\begin{aligned} M &: \ a + b = 0 \\ L &: \ -a - b + 2c = 0 \\ T &: -2a - 2b - c + e = 0 \end{aligned}$$

The solution of this system of equations gives

$$a + b, \ c = e, \ d = -2e$$

where b and e are arbitrary constants. If we set $b = 1$and $e = 0$, we obtain $a = -1$, $c = 0$, and $d = 0$; likewise, $b = 0, e = 1$ which produces $a = 0, c = 1$, and $d = -2$. These independent solutions yield the complete set of dimensionless products:

$$\Pi_1 = \Delta T_{\mathrm{m}}^{-1} \Delta T_{\mathrm{c}} \quad \text{and} \quad \Pi_2 = k l^{-2} t_{\mathrm{c}}$$

From Buckingham's theorem, we obtain the dependency

$$h(\Pi_1, \Pi_2) = 0$$

or

$$t_{\mathrm{c}} = \left(\frac{l^2}{k}\right) H\left(\frac{\Delta T_{\mathrm{c}}}{\Delta T_{\mathrm{m}}}\right) \tag{2}$$

The rule stated at this example gives the roasting time for the turkey in terms of its weight w. Let us assume all the turkeys are geometrically similar and have volume of V that can be written as characteristic length of l in the form of $V \propto l^3$. If we assume the turkey is of constant density, which in reality is not quite the case because of the bones and flesh of each turkey from the other one in density. Since the weight is density-multiplying volume and volume is proportional to l^3, we get $w \propto l^3$. Moreover, if we set the oven to a constant baking temperature and specify that the turkey must initially be near room temperature of 65 °F, then $\Delta T_c / \Delta T_{\mathrm{m}}$ is a dimensionless constant. Combining these results with Eq. 2 above, we get the proportionality

$$t \propto w^{2/3} \tag{3}$$

because the thermal conductivity k is constant for all turkeys. Thus, the required cooking time is proportional to weight raised to the two-thirds power. Therefore, if t_1 hours are required to cook a turkey weighing w_1 pounds and t_2 is the time for a weight w_2 pounds

$$\frac{t_1}{t_2} = \left(\frac{w_1}{w_2}\right)^{2/3}$$

it follows that a doubling of the weight of a turkey increases the cooking time by the factor $2^{2/3} \approx 1.59$ (Figs. 1.8 and 1.9).

How does our result of Eq. 3 above compare to the rule stated previously? Assume that ΔT_{m}, ΔT_{c}, and k are independent of the length or weight of the turkey and consider cooking a 23-lb turkey versus and 8-lb bird. According to our rule, the ratio of cooking times is given by

$$\frac{t_1}{t_2} = \left(\frac{23}{8}\right)^{2/3} \approx 2.02$$

Fig. 1.8 Schematic of a very large turkey of a characteristic length l along with cookbook data for the cooking time t_c as a function of the mass of the turkey m

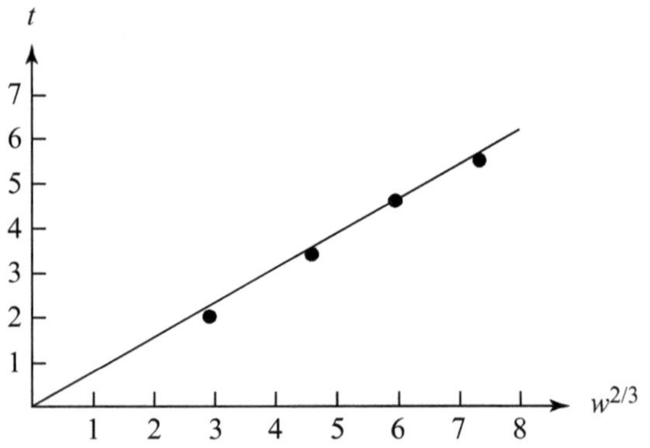

Fig. 1.9 Plot of cooking times versus weight to the two-thirds power reveals the predicted proportionality

Thus, the rule predicts it will take nearly three times as long to cook a 23-lb bird as it will to cook an 8-lb turkey. Dimensional analysis predicts it will take only twice as long. The question then is which rule is correct? Why there are so many cooks overcook a turkey? The testing results can be reasoned as follows: Suppose that turkeys of various sizes are cooked in an oven preheated to 325 °F. The initial temperature of the turkey is 65 °F. All the turkeys are removed from the oven when their internal temperature, measured by a meat thermometer, reaches 195 °F. The hypothetical cooking times for the various turkeys are recorded as shown in the above table in Fig. 1.8, which is presented as a plot of t versus $w^{2/3}$. Because the graph approximates a straight line through the origin, we conclude that $t \propto w^{2/3}$, as predicted by our model.

As part of the turkey roasting problem, we have showed another approach, and we leave it to reader to make his or her own choice of correctness of this example.

We use this approach as a second option of solving roasting turkey problem the way John Ruebush of University of Utah, Salt Lake City, UT 84105 and Robert Fisk of Colorado School of Mines, Golden, Colorado was showing. The Mathematical Association of America is collaborating with JSTOR Vol. 53, NO. 4, September 1980.

Many cookbooks provide a table of cooking time versus weight for roasting meats of various types. For roasting a turkey, the weight-time information in Table 1 is provided by the *Betty Crocker Cookbook* ([1], p. 442). We shall use some dimensional analysis and scale modeling to attempt to verify the cooking time entries in Table 1, assuming a cooking time of 3.25 h for a 7-lb turkey.

Table 1 Betty cookbook data

Ready-to-cook weight (lb)	Approximate cooking time (h)	Internal temperature (°F)
6–8	3.00–3.25	1850
8–12	3.25–4.25	1850
12–16	4.25–5.25	1850
16–20	5.25–6.25	1850
20–24	6.25–7.00	1850

We start with the one-dimensional diffusion equation

$$\frac{\partial \theta}{\partial t} = \kappa \frac{\partial^2 \theta}{\partial x^2} \tag{1}$$

where $\theta(x,t)$ represents the temperature at any point x and any instant t, assuming a homogeneous "turkey" with κ the coefficient of thermal diffusivity. Let us assume the turkey is initially at room temperature T_1 and that the oven is at temperature T_2; we put the turkey in the oven at time $t = 0$. Let L be the length of the turkey. Then we have the boundary and initial conditions:

$$\begin{aligned} \theta(x,0) &= T_1 \quad 0 < x < L \\ \theta(0,t) &= T_2 \quad t < 0 \\ \theta(L,t) &= T_2 \quad t < 0 \end{aligned} \tag{2}$$

Next, we write Eqs. 1 and 2 in a dimensionless form by letting (see [2])

$$\xi = \frac{x}{L} \quad \tau = \frac{\kappa t}{L^2} \quad \phi = \frac{\theta - T_1}{T_2 - T_1} \tag{3}$$

First we find $\partial\theta/\partial t$ and $\partial^2\theta/\partial x^2$ in terms of ξ, τ, and ϕ. Since

(continued)

(continued)

$$\frac{\partial\theta}{\partial t} = \frac{d\theta}{d\phi}\cdot\frac{\partial\phi}{\partial\tau}\cdot\frac{\partial\tau}{dt} = (T_2 - T_1)\cdot\frac{\partial\phi}{\partial\tau}\cdot\frac{\kappa}{L^2}$$

we have

$$\frac{\partial\theta}{\partial t} = \frac{(T_2 - T_1)\kappa}{L^2}\frac{\partial\phi}{\partial\tau} \tag{4}$$

Similarly, since

$$\frac{\partial\theta}{\partial x} = \frac{d\theta}{d\phi}\cdot\frac{\partial\phi}{\partial\xi}\cdot\frac{d\xi}{dx} = (T_2 - T_1)\cdot\frac{\partial\phi}{\partial\xi}\cdot\frac{1}{L}$$

we obtain

$$\frac{\partial\theta}{\partial x} = \frac{T_2 - T_1}{L}\frac{\partial\phi}{\partial\xi}$$

Hence

$$\begin{aligned}\frac{\partial^2\theta}{\partial x^2} = \frac{\partial}{\partial x}\left(\frac{T_2 - T_1}{L}\cdot\frac{\partial\phi}{\partial\xi}\right) &= \frac{T_2 - T_1}{L}\cdot\frac{\partial}{\partial x}\left(\frac{\partial\phi}{\partial\xi}\right)\\ &= \frac{T_2 - T_1}{L}\cdot\frac{\partial}{\partial\xi}\left(\frac{\partial\phi}{\partial\xi}\right)\cdot\frac{d\xi}{dx}\\ &= \frac{T_2 - T_1}{L}\cdot\frac{\partial^2\phi}{\partial\xi^2}\cdot\frac{1}{L} = \frac{T_2 - T_1}{L^2}\cdot\frac{\partial^2\phi}{\partial\xi^2}\end{aligned} \tag{5}$$

Using Eqs. 4 and 5 in Eq. 1 yields

$$\frac{\partial\phi}{\partial\tau} = \frac{\partial^2\phi}{\partial\xi^2} \tag{6}$$

and using Eq. 3 in Eq. 2 gives

$$\begin{aligned}&\phi(\xi, 0) = 0 \quad 0 < \xi < 1\\ &\phi(0, \tau) = 1 \quad \tau > 0\\ &\phi(1, \tau) = 1 \quad \tau > 0\end{aligned} \tag{7}$$

The importance of Eq. 6 is that the solution in the dimensionless form does not depend on what is being heated and, in particular, does not depend on the size of the object being heated.

(continued)

(continued)

In the problem of roasting turkeys of different sizes, we wish to achieve the same internal temperature for each bird. This takes some time τ_0 in the dimensionless form. We assume that turkeys of different sizes (weights) are geometrically similar objects and that they have the same thermal characteristics (i.e., κ is the same for all turkeys). We also assume that the weight of a turkey is proportional to the cube of its length (see [3]):

$$W^{1/3} = kL \tag{8}$$

Now, consider two turkeys, 1 and 2, which have weights W_1 and W_2, respectively. We assume that each is to be roasted until its temperature is T. In the nondimensionless form, this takes times t_1 and t_2, respectively, for each turkey. But, in the dimensionless form, τ_1 equals τ_2, so from Eq. 3

$$\frac{\kappa t_1}{L_1^2} = \frac{\kappa t_2}{L_2^2} \tag{9}$$

where t_1 and t_2 are the cooking times for the two turkeys and L_1 and L_2 are the lengths of the birds. Using Eq. 8 in Eq. 9 leads to

$$t_2 = \left(\frac{W_2}{W_1}\right)^{2/3} t_1 \tag{10}$$

This says that cooking times are not linear with weight ratios, but go as the 2/3 power.

Table 2 Cooking time

Weight (lb)	Time (h)
10	4.1
14	5.2
18	6.1
22	7.0

If we take $t_1 = 3.25$ hours for a 7-lb turkey, then we get the cooking times in Table 2, using Eq. 10 for turkeys of other weights. It is clear from this table that our simple model confirms the cooking table provided by Betty Crocker. It is interesting to observe that we avoided the solution of the heat equation in getting the cooking times.

References

1. Betty Crocker's Cookbook, Bantam Books, New York, pp. 442.
2. C. C. Lin and L. A. Segel, Mathematics Applied to Deterministic Problems in the Natural Sciences, Macmillan, New York, 1974, p. 195.
3. J. Maynard Smith, Mathematical Ideas in Biology, Cambridge Univ. Press, London, 1968, pp. 2–13.

1.10 Energy in a Nuclear Explosion

We next turn to a famous example worked out by the eminent British physicist Sir G. I. Taylor (1886–1975). When the United States exploded the first nuclear bomb under code name "Trinity" as part of Manhattan Project at a location about 100 miles south of Los Alamos in the aptly named Jornada del Muerto (Journey of Death) in the Alamogordo Desert of southern New Mexico in 1945, the site was selected to test the implosion device. Oppenheimer named the site "Trinity" after a poem that he had been reading.

The event of this explosion and everything associated with it were under shroud of secrecy, and both data and the motion picture of explosion footage from the explosion were highly classified. Two years later, the US government released a series of pictures of the explosion, along with a size scale and time stamps of the explosion (see Figs. 1.10 and 1.11 Trinity explosion and its sequence). G. I. Taylor managed to determine the energy released during the explosion using only the expansion radius of blast wave at time t and the principle of dimension analysis. Later on, when data were declassified, his analysis and calculations were remarkably accurate to what these data were showing.

Taylor first treated the problem in 1941, with a numerical solution for an explosion in air. Taylor's initially classified paper on the subject was published in 1949 as "The Formation of a Blast Wave by a Very Intense Explosion; I [20] Theoretical Discussion." The theoretical discussion of 1941 included some fairly crude experimental comparisons, based on the blasts of conventional weapons, and Taylor also added some better conventional weapons data for the 1949 publication.

As a better experimental test, Taylor produced a companion article, "The Formation of a Blast Wave by a Very Intense Explosion; II [21] The Atomic Explosion of 1945." Using a time series of then recently declassified photographs of the Trinity explosion near Alamogordo, New Mexico, Taylor tested his scaling

Fig. 1.10 Trinity explosion after $t = 0.025$ s

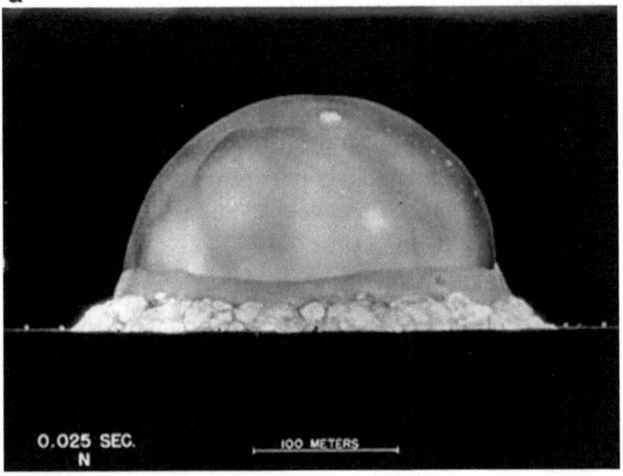

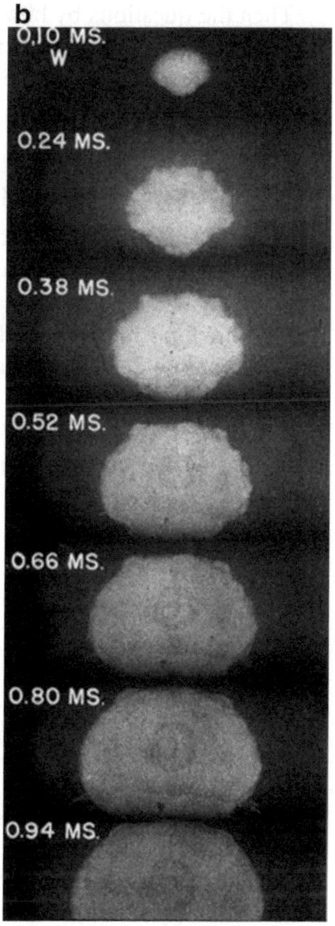

Fig. 1.11 Trinity explosion sequence

hypothesis and calculated the energy of the blast. The frequently quoted strength of the explosion—18.6 kt (the equivalent of 18.6 thousand tons of trinitrotoluene)—is quite close to Taylor's calculated value of 16.8 kt.

While Taylor's scaling law was correct, and his numerical solution worked remarkably well for the case of air, a more general solution was, needed, to study a blast wave propagating into a cold gas with a more general equation of state. This is important, for example, in the study of supernova explosions. This need had been already answered by the time Taylor published his paper in 1949. Sedov and von Neumann independently found an analytic solution to the problem, for the same set of approximations that Taylor used. For this reason, the solution to the blast problem is conventionally known as the "Sedov solution."

Basically for his calculation, first, he needed two assumptions as follows:

1. The energy E was released in a small space.
2. The shock wave was spherical.

Then the questions by him were that:

We have the size of the fireball radius (R) as a function of time (t) at several different times. How does the radius (R) depend on:

- Energy (E)
- Time (t)
- Density of the surrounding medium (ρ_0 initial density of air)

In a nuclear explosion, there is an essentially instantaneous release of energy E in a small region of space. This produces a spherical shock wave, with the pressure inside the shock wave thousands of times greater than the initial air pressure, which may be neglected. How does the radius R of this shock wave grow with time t? The relevant governing variables and $[R] = L$ as function of these variable are E, t, and the initial air density ρ_0, with dimensions $[E] = ML^2T^{-2}$, $[t] = T$, and $[\rho_0] = ML^{-3}$. This set of variables has independent dimensions, so $n = 3$ and $k = 3$. We next determine the exponents by substituting the

$$\begin{aligned} R &= f(E, \rho_0, t) \\ [R] = L &= [E]^a [\rho_0]^b [t]^c \\ &= \left[ML^2T^{-2}\right]^a \left[ML^{-3}\right]^b [T]^c \end{aligned}$$

or

$$L = M^{a+b} L^{2a-3b} T^{-2a+c}$$

M is to the $a + b$ power because energy and density are both dependent on M. L is to the $2a - 3b$ power because energy is dependent on the square of distance and density is dependent on one over the cube of distance. T is to the $-2a + c$ power because energy is dependent on one over the square of time and time is dependent on time. Therefore, this provides three simultaneous equations as follows:

$$a + b = 0, \quad 2a - 3b = 1, \quad -2a + c = 0$$

with solution $a = 1/5$, $b = -1/5$ and $c = 2/5$. Therefore, the radius R is given by Eq. 1.21 as follows:

$$R = CE^{1/5} \rho_0^{-1/5} t^{2/5} \tag{1.21}$$

with C being an undetermined constant. If we could plot the radius R of the shock as a function of time t on a log–log plot, the slope of the line should be 2/5. The intercept of the graph would provide information about the energy E released in the explosion, if the constant C could be determined. By solving a model shock-wave problem Taylor estimated C to be close to unity ($C = 1.033$ for adiabatic index of air $\gamma = 1.4$); he was able to take declassified movies of nuclear tests, and using his

Table 1.6 Taylor's original data with the time t measured in milliseconds and the radius R in meter [24]

t	$R(t)$	t	$R(t)$	t	$R(t)$	t	$R(t)$	t	$R(t)$
0.10	11.1	0.80	34.2	1.50	44.4	3.53	61.1	15.0	106.5
0.24	19.9	0.94	36.3	1.65	46.0	3.80	62.9	25.0	130.0
0.38	25.4	1.08	38.9	1.79	46.9	4.07	64.3	34.0	145.0
0.52	28.8	1.22	41.0	1.93	48.7	4.34	65.6	53.0	175.0
0.66	31.9	1.36	42.8	3.26	59.0	4.61	67.3	62.0	185.0

model, infer the yield of the bombs. The final version of his analysis and calculation is summarized in form of the following expression (Eq. 1.22):

$$R = E^{1/5}\rho_0^{-1/5}t^{2/5} \tag{1.22}$$

Again we have assumed the constant C is approximately 1 ($C \approx 1$), and this implies that, to a good approximation, radius of shock wave R can be presented by the above equation.

We can now convert to a linear function in order to compute the energy. Taking the logarithm of both sides produces

$$\begin{cases} \log_{10}R = \dfrac{2}{5}\log_{10}t + \dfrac{1}{5}\log_{10}\left(\dfrac{E}{\rho_0}\right) \\ \dfrac{5}{2}\log_{10}R = \log_{10}t + \dfrac{1}{2}\log_{10}\left(\dfrac{E}{\rho_0}\right) \end{cases}$$

where $\rho_0 = 1.2\ \text{kg/m}^3$. Figures 1.10 and 1.11 illustrate the transformed data from Taylor's original values given in Table 1.6. A least-square fit of this data gives an estimate of $(1/2)\log_{10}(E/\rho_0) \simeq 6.90$ so that we have $E = 8.05 \times 10^{13}$. Using the conversion factor of 1 kt = 4.168×10^{12} J gives the strength of Trinity as 19.2 kt. It was later revealed that the actual strength of Trinity explosion was 21 kt. This demonstrates the predictive power of dimensional analysis.

The Method of Least Squares

The method of least squares assumes that the best-fitting curve of a given type is the curve that has the minimal sum of the deviations squared (least square error) from a given set of data.

Suppose that the data points are $(x_1, y_1), (x_2, y_2), \ldots, (x_n, y_n)$ where x is the independent variable and y is the dependent variable. The fitting curve $f(x)$ has the deviation (error) $E \approx 21$ from each data point, i.e., $d_1 = y_1 - f(x_1)$, $d_2 = y_2 - f(x_2)$, ..., $d_n = y_n - f(x_n)$. According to the method of least squares, the best-fitting curve has the property that (Fig. 1.12)

(continued)

(continued)

$$\Pi = d_1^2 + d_2^2 + \cdots + d_n^2 = \sum_{i=1}^{n} d_i^2 = \sum_{i=1}^{n} [y_i - f(x_i)]^2 = \text{a minimum}$$

Solving the equation for E we get

$$E = \left(R^5 \rho_0\right)/t^2$$

Looking at Fig. 1.10 and using the number on the slide, we see that at $t = 0.025$ seconds the radius of the shock-wave front was approximately 100 m. The density of air ρ_0 typically is $\rho_0 = 1.2\ \text{kg/m}^3$. Plugging these values into the energy equation gives

$$\begin{aligned} E &= (100)^5 \times 1.2/(0.025)^2 [\text{kg m}^2/\text{s}^2] \\ &= 1.92 \times 10^{10} [\text{kg m}^2/\text{s}^2] \\ &= 1.92 \times 10^{17} \text{ergs} \end{aligned}$$

Now, we know that 1 g of TNT is equal to 4×10^{10} ergs, and hence

$$E \approx 21 \quad \text{Kilotons of TNT}$$

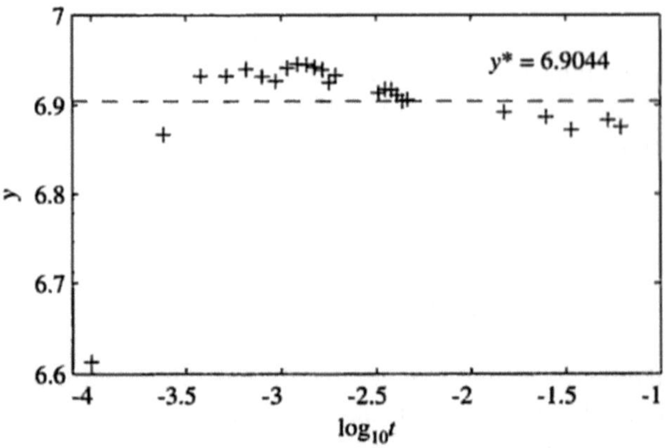

Fig. 1.12 The graph of $Y = 5/2\log_{10}r - \log_{10}t$ is identified with the quantity $1/2\log_{10}(E/\rho_0)$. A least square fit yields $y^* = 6.90$, and because ρ_0 is known, it is possible to solve for E, the energy released during the explosion [24]

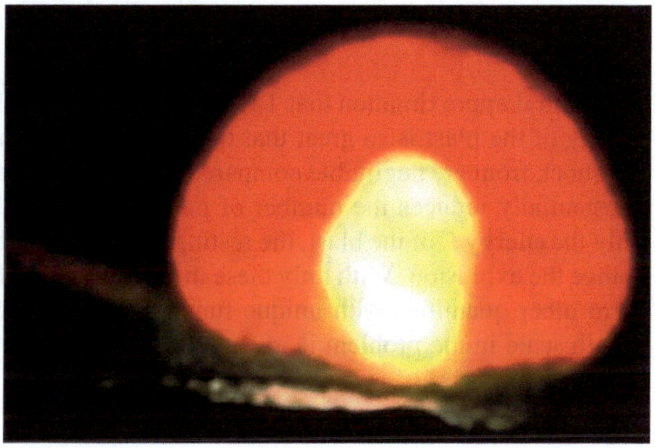

Fig. 1.13 Teak photographed from Hawaii, 1 min after detonation 1300 km away, 1 August 1958, a 3.8 Mt burst at 77 km directly over Johnston Island

This data, of course, was strictly classified; it came as a surprise to the American intelligence communities that this data were essentially publicly available to those well versed in dimensional analysis. Later, G. I. Taylor [20, 21] processing the photographs taken by H. E. Mack (Figs. 1.10 and 1.11) of the first atomic explosion in New Mexico in July 1945 confirmed this scaling law a well-deserved triumph of Taylor intuition.

It is important to note one further fact. A nuclear explosion has several unique properties; Taylor's paper is a description of only one. The first (and ultimately, most deadly) part of the explosion is, the blinding flash, the burst of electromagnetic radiation released by the nuclear processes (either fission or fusion), Fig. 1.13. The second stage is the "fireball," the beautifully spherical shock front that the Sedov solution describes. Finally, there is the infamous mushroom cloud, which comes about through the interaction of the fireball interior with the ground and the stratification of the atmosphere. Also other, more subtle effects are of little importance in calculating the damaging power of a bomb, but are physically interesting nonetheless. For example, gamma rays may ionize the atmosphere outside of the fireball, surrounding the sphere with a penumbra of lightning strikes.

Note: On the left-hand side, a faint aurora from the base of the radioactive fireball shows the path of radiation traveling along the Earth's magnetic field toward the southern hemisphere.

Glasstone and Dolan explain:

> Because the primary thermal radiation energy in a high-altitude burst is deposited in a much larger volume of air, the energy per unit volume available for the development of the shock front is less than in an air burst. The air at the shock front does not become hot enough to be opaque. There is no apparent temperature minimum as is the case for an airburst. The thermal radiation is emitted in a single pulse. The reason is that formation of ozone, oxides of nitrogen, and nitrous acid, which absorb strongly in this spectral region, is decreased [because these chemicals, which give the fireball its rust-like color and create the thermal minimum by absorbing fireball radiation until the shock front breaks away from the fireball and cools, only form in a dense, compressed shock wave].

1.10.1 The Basic Scaling Argument in a Nuclear Explosion

The central approximation that Taylor, Sedov, and von Neumann use is simple. The energy of the blast is so great that the pressure and temperature of the gas outside the shock front are negligible compared to the pressure and temperature inside. This substantially reduces the number of parameters available in the problem, leaving only the energy E of the blast, the resting density ρ_0 of the external gas, and the time t since the explosion. With only these three dimensional parameters, it is possible to form other quantities with unique functional dependences. In particular, the only length scale in the problem is given by Eq. 1.23 as

$$R \propto C(\gamma) E^{1/5} \rho_0^{-1/5} t^{2/5} \tag{1.23}$$

The constant of proportionality $C(\gamma)$ will depend on the equation of state of the gas and for ideal gas is roughly 1 and for air is about 1.4.

The radius R of the fireball forming the outer edge of the disturbance is the simplest quantity to measure, but to solve the problem in detail, one must introduce the scaled forms of pressure, density, and radial velocity in the gas. These will naturally be functions of distance r from the center of the blast in radial coordinate. However, since R is the only length scale at a given time, the dependence must be through a function of the dimensionless ratio $\eta \equiv \frac{r}{R}$. Taylor introduces functions f_1, y, ψ, and ϕ_1 of η, so that the physical quantities varied according to Eq. 1.24. This way Taylor was to demonstrate in the same way the scaling laws for pressure, velocity, and density immediately behind the shock-wave front of the blast:

$$\begin{aligned} &\text{Pressure} \quad p/p_0 = y = R^{-3} f_1 \\ &\text{Density} \quad \rho/\rho_0 = \psi \\ &\text{Radial Velocity} \quad u = R^{-3/2} \phi_1 \end{aligned} \tag{1.24}$$

where p is the pressure front (with p_0 equal to its resting value or initial undisturbed ambient air pressure), ρ is the front air density (with ρ_0 defined as initial density in ambient air accordingly), and u is the radial velocity of the gas expanding from blast center. In radial coordinate r, $\eta = \frac{r}{R}$ and f_1, ϕ_1, and ψ are functions of η. The set of Eq. 1.24 were established as an appropriate similarity assumption for an expanding blast wave of constant total energy. With these functions, the solution can be found by straightforward algebra since it is found that these assumptions are consistent with the equations of motion, continuity equation as well as equation of state for a perfect gas [20]. Taylor assumed several important crucial points that allowed him to obtain the solution of set of partial differential equation such as set of equation of motion, (i.e., *conservation of energy* and *conservation of momentum and conservation of energy equations*) for expanding shock front which was needed in a simple and effective form.

Note that for self-similar motions, the system of partial differential equations of gas dynamics (in this case expanding shock front from the intense blast) reduces to a system of ordinary differential equations in new unknown functions of similarity variable $\eta = \frac{r}{R}$. This reduction of partial differential equations to ordinary differential equations induced from the fact if we consider one-dimensional adiabatic flows of a perfect gas with constant specific heats, with planar, cylindrical, or spherical symmetry. Then the system of equations representing this type of flow can be written [25] as follows (Eq. 1.25):

$$\begin{gathered}\frac{\partial \ln\rho}{\partial t}+u\frac{\partial \ln\rho}{\partial r}+\frac{\partial u}{\partial r}+(\nu-1)\frac{u}{r}=0\\ \frac{\partial u}{\partial t}+u\frac{\partial u}{\partial r}+\frac{1}{\rho}\frac{\partial p}{\partial r}=0\\ \frac{\partial}{\partial t}\ln p\rho^{-r}+u\frac{\partial}{\partial r}\ln p\rho^{-r}=0.\end{gathered} \tag{1.25}$$

Derivation of Eq. 1.25

In the following continuity equation which describes the conservation of mass of the fluid and in fact that is density in a given volume element changes as a result of flow of the fluid into or out of this element:

$$\frac{\partial \rho}{\partial t}+\vec{\nabla}\cdot\rho\vec{u}=0$$

In this equation we expand the divergence term and write the equation in a form appropriate to all three types of symmetry. In addition, we divide the equation through by the density ρ. The entropy in the following entropy equation

$$\frac{DS}{Dt}=0$$

which is expressed in different format with the density in place of specific volume for a perfect gas with constant specific heats can be expressed in a simple form in terms of pressure and density (in specific volume):

$$S=c_p\ln pV^{\gamma}+\text{constant}$$

where S is the specific entropy and is the isentropic exponent, equal to the ration of the specific heats at constant pressure and at constant volume,

(continued)

(continued)

Derivation of Eq. 1.25

$\gamma = c_p/c_v = 1 - R/c_p$, $\vec{u}$ being the velocity vector, and R is the gas constant per unit mass. Note that a partial derivative with respect to time at a given point in space is denoted by $\partial/\partial t$, and a total derivative, describing the time change in any quantity following a moving fluid particle, by D/Dt.

The equation motion that we also need is expressed for an incompressible fluid, where $\rho =$ constant, and the continuity equation is $\vec{\nabla} \cdot \vec{u} = 0$. The equation of motion expresses Newton's law and does not differ from the corresponding equation of motion for an incompressible fluid (p is the pressure):

$$\rho \frac{D\vec{u}}{Dt} = -\nabla p$$

or, in the form of Euler's equation, can be written as

$$\frac{\partial \vec{u}}{\partial t} + \vec{u} \cdot \nabla \vec{u} = -\frac{1}{\rho} \nabla p$$

The above equation can be left unchanged.

In Eq. 1.25 for the continuity $\nu = 1$, 2, and 3 for the plane, cylindrical, and spherical cases, respectively. The variable r represents the x coordinate in the plane case and the radius in the cylindrical and spherical cases.

The initial conditions for solving these set of equation at the beginning stage of the very intense shock propagating due to the nuclear blast were describe as follows:

$$\rho(r,0) = \rho_0 \quad p(r,0) = p_0 \quad u(r,0) = 0 \quad \text{for} \quad r \geq r_0 \tag{1.26}$$

Here r_0 is the initial radius of the shock wave that outstrips the thermal wave. The steps that he took and were an ideal on as he later published in his papers [20, 21] are defined here:

"A finite amount of energy is suddenly released in an infinitely concentrated form." This assumes that the explosion coming from a point source of energy so consequently r_0 is taken to be zero. This assumption then clearly indicates neglecting the initial radius r_0 of the shock wave is allowed only when the motion is considered at a stage when the shock front radius r is much larger than r_0 as it can be seen in Eq. 1.26. If the initial shock-wave radius is taken equal to zero, then the initial distributions of the air density, pressure, and velocity inside the initial shock wave disappear from the problem statement that Taylor was facing [10]. At the same time, Taylor restricted himself to consider the motion of the shock front when

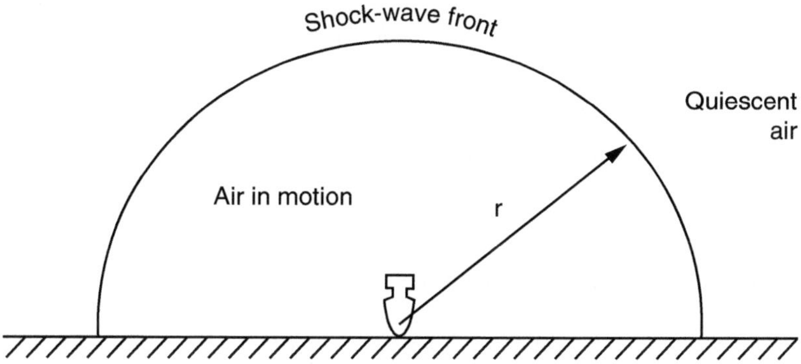

Fig. 1.14 Very intense shock-wave propagation in quiescent ambient air

the maximum pressure of the moving expanded gas is much larger than the pressure p_0 in the ambient gas. This assumption then will allow neglecting the terms involving the initial pressure p_0 in the boundary conditions at the shock wave that are associated with the equation of motion inside the shock. He assumed the motion for highly intense point explosion in ambient quiescent air expanding outward is spherically symmetric fashion and radii r going out from the explosion center point are identical in every direction as depicted in Fig. 1.14. Incidentally such a simplified assumption by him pretty much was confirmed by first atomic explosion, which is shown in Fig. 1.10.

1.10.2 Calculating the Differential Equations of Expanding Gas of Nuclear Explosion

Taylor went on to calculate the set of differential equations that he needed to carry on the analysis for such intense explosion in his famous published paper "The Formation of a Blast Wave by a Very Intense Explosion." Hence the following equation of motion inside the shock wave had to be considered by writing the basic form of such equation for expanding gas as follows:

$$u_t + uu_r = -\frac{p_0}{\rho} y_r \tag{1.27}$$

With the self-similar Eq. 1.23, and substituting the entire set of relationship in Eq. 1.24, and like he says in his paper writing $f_1' = \frac{\partial f_1}{\partial \eta}$ and $\phi_1' = \frac{\partial \phi_1}{\partial \eta}$, the Eq. 1.27 becomes

$$-\left(\frac{3}{2}\phi_1+\eta\phi_1'\right)R^{-\frac{5}{2}}\frac{dR}{dt}+R^{-4}\left(\phi_1\phi_1'+\frac{p_0 f_1'}{\rho_0\psi}\right)=0 \tag{1.28}$$

In keeping with the expected scaling law, Taylor lets

$$\frac{dR}{dt}=U=AR^{-\frac{3}{2}} \tag{1.29}$$

which gives an equation for constant A:

$$A\left(\frac{3}{2}+\eta\phi_1'\right)+\phi_1\phi_1'+\frac{p_0}{\rho_0}\frac{f_1'}{\psi}=0 \tag{1.30}$$

Equation 1.30 is the first of the final equations of motion.

The second equation arises from the continuity equation, expressing mass conservation as follows:

$$\frac{\partial\rho}{\partial t}+u\frac{\partial\rho}{\partial r}+\rho\left(\frac{\partial u}{\partial r}+\frac{2u}{r}\right)=0 \tag{1.31}$$

Substituting Eq. 1.24 set along with Eq. 1.29 into Eq. 1.31, then we have

$$-A\eta\psi'+\psi'\phi_1+\psi\left(\phi_1'+\frac{2}{\eta}\phi_1\right)=0 \tag{1.32}$$

The final equation derives from the equation of state. At this point, a further approximation is needed; the gas must have a polytropic equation of state, $p=\kappa\rho^\gamma$. This approximation is often very accurate. The number γ (also known as adiabatic index) is the ratio of specific heats of the gas, C_P/C_V, and it represents the number n of degrees of freedom available, according to $\gamma=\frac{n+2}{n}$. For a monoatomic gas, there are only the three translational degrees of freedom (i.e., $n = 3$), so $\gamma=\frac{5}{3}$. For a diatomic gas, there are two additional rotation axes with nonzero moments of inertia, so $n = 5$, and $\gamma=\frac{7}{3}$. For air, which is 99 % diatomic (with N^2 and O^2), the measured value of γ is 1.40. However, there are other values of interest. The interstellar medium contains mostly atomic hydrogen gas, H^1, so the monoatomic value of γ is more useful there. (In fact, Taylor did numerical calculations for both these values, as well as for $\gamma = 1.20$ and $\gamma = 1.30$.)

For whatever value of γ is appropriate, the equation of state may be written in the alternative form that Taylor uses:

$$\left(\frac{\partial}{\partial t}+u\frac{\partial}{\partial r}\right)(p\rho^{-\gamma})=0 \tag{1.33}$$

With the equations above, this can be written in the (rather more complicated) form which easily can be derived by substituting set of Eqs. 1.14 and 1.30 into Eq. 1.33, and then it yields

$$A(3f_1 + \eta f_1') + \frac{rf_1}{\psi}\psi'(-A\eta + \phi_1) - \phi_1 f_1' = 0 \tag{1.34}$$

This is the last of the three dynamical equations.

Taylor's last step in analyzing these equations is to nondimensionalize f_1 and ϕ_1, using the resting sound speed $a = (\gamma p_0/\rho_0)^{1/2}$. The dimensionless functions f and ϕ are then given by Eqs. 1.35 and 1.36 as follows:

$$f = \frac{a^2}{A^2} f_1 \tag{1.35}$$

$$\phi = \frac{1}{A}\phi_1 \tag{1.36}$$

This substitution eliminates A from the equations of motion. (If A did not fall out, it would imply the existence of some other dimensional quantity in the problem.) The nondimensional equations of motion finally become

$$\begin{aligned} &\phi'(\eta - \phi) - \frac{1}{\gamma} f' \psi^{-1} - \frac{3}{2}\phi = 0 \\ &\psi'\psi^{-1} - \frac{\phi' + 2\phi\eta^{-1}}{\eta - \phi} = 0 \\ &3f + \eta f' + \gamma\psi'\psi^{-1} f(\phi - \eta) - \phi f' = 0 \end{aligned} \tag{1.37}$$

This system of differential equations can be solved by an ordinary procedure, stepping through values of η in either direction. Taylor [20] shows this step by eliminating ψ from the third equation above by means of using the other two sets in Eq. 1.37 and obtains the following form:

$$f'\left\{(\eta - \phi)^2 - \frac{f}{\psi}\right\} = f\left\{-3\eta + \phi\left(3 + \frac{1}{2}\gamma\right) - 2\gamma\frac{\phi^2}{\eta}\right\} \tag{1.38}$$

When f' has been found from Eq. 1.38, then the other two variables, ϕ' and ψ', can be calculated from the other first two sets of Eq. 1.37. Thus, if for any values of η, f, ϕ, and ψ are known, their values can be completed step by step for other values of η as mentioned above.

1.10.3 Solving the Differential Equations of Expanding Gas of Nuclear Explosion

To solve the differential equations still requires initial conditions. However, these are easy to find. Since Rayleigh and Taylor's pioneering work on shocks, general shock conditions expressing conservation of mass, momentum, and energy had been formulated. Of course, as Taylor notes, the general conditions are inconsistent with the basic similarity hypothesis of Eq. 1.23. That is perfectly natural. The completely generic shock conditions include allowances for the character of the gas downstream from the shock. The similarity hypothesis requires that these conditions be essentially irrelevant or that one may neglect factors of order unity compared to values of y inside the fireball.

Taylor denotes the value of a quantity $\eta = 1$ immediately inside the shock and uses Rankine–Hugoniot relations, which may be reduced to the following form:

$$\begin{aligned}\frac{\rho_1}{\rho_0} &= \frac{\gamma - 1 + (\gamma + 1)y_1}{\gamma + 1 + (\gamma - 1)y_1} \\ \frac{U^2}{a^2} &= \frac{1}{2\gamma}\{\gamma - 1 + (\gamma + 1)y_1\} \\ \frac{u_1}{U} &= \frac{2(y_1 - 1)}{\gamma - 1 + (\gamma + 1)y_1}\end{aligned} \tag{1.39}$$

where ρ_1, u_1, and y_1 represent the values of ρ, u, and y immediately behind the shock wave and $U = dR/dT$ is the radial velocity of the shock wave (fireball). With this notation, and using the boundary conditions for $y_1 \gg 1$, the limit for set of Eq. 1.39 then are presented in a new form set such as Eq. 1.40:

$$\begin{aligned}\frac{\rho_1}{\rho_0} &= \frac{\gamma - 1}{\gamma + 1} \\ \frac{U^2}{a^2} &= \frac{2\gamma}{\gamma + 1}y_1 \\ \frac{u_1}{U} &= \frac{2}{\gamma + 1}\end{aligned} \tag{1.40}$$

Note that the shock wave moves faster than the gas behind it, so that the shock wave does not carry any material along with it. In terms of the dimensionless functions, the boundary conditions in Eq. 1.33 approximation for $\eta = 1$ are inconsistent with Eqs. 1.24 and 1.29, where in fact the first set of Eq. 1.41 yields

$$\begin{aligned}\psi &= \frac{\gamma - 1}{\gamma + 1} \\ f &= \frac{2\gamma}{\gamma + 1} \\ \phi &= \frac{2}{\gamma + 1}\end{aligned} \tag{1.41}$$

Taylor solves these equations numerically, for comparison with conventional high explosives and with the Trinity pictures. Most of the pair of papers are devoted to this endeavor. For the conventional explosives, it was really too difficult to measure the size of the actual blast wave. Taylor instead studies the residual effects, such as the heating that remains after the passing of the blast.

Measurements of this sort are rather suspected, because they should be affected by the presence of the ground. The blast striking the ground may produce a rebound shock wave. However, this shock front lies entirely inside the fireball. So the radius of the fireball is unaffected by the ground, but the interior conditions are not. In fact, Taylor's predictions do not agree very well with the conventional explosion data on many points.

For the nuclear explosion, Taylor's data is actually quite good. In this case, the blast radius is easy to measure, and the strength of the shock really is astronomical compared to external conditions. Included with this summary are three of the pictures from the time sequence that Taylor used. The folklore of this problem makes a big deal of Taylor's calculation of the energy of the blast, and this is the last major theoretical calculation that he does.

The energy has two parts, the bulk kinetic energy of the motion and the thermal energy of the gas. The expressions for them are elementary integrals:

$$\begin{aligned} K.E. &= 4\pi\int_0^R \frac{1}{2}\rho u^2 r^2 dr \\ H.E. &= 4\pi\int_0^R \frac{pr^2}{\gamma-1} dr \end{aligned} \tag{1.42}$$

Taylor then notes that by pulling out a factor of $\rho_0 A^2$, he gets two dimensionless integrals in terms of the variables f, ϕ, ψ, and η:

$$E = \rho_0 A^2\left\{\frac{1}{2}\rho_0\int_0^1 \psi\phi^2\eta^2 d\eta + \left(\frac{p_0}{a^2(\gamma-1)}\int_0^1 f\eta^2 d\eta\right)\right\}$$

or since $p_0 = a^2\rho_0/\gamma$ and $E = B\rho_0 A^2$, where a function of only whose value is give in form Eq. 1.43 as follows;

$$B = 2\pi\int_0^1 \psi\phi^2\eta^2 d\eta + \frac{4\pi}{\gamma(\gamma-1)}\int_0^1 f\eta^2 d\eta \tag{1.43}$$

Since the two integrals in Eq. 1.43 are both functions of γ, only it seems that for a given value of γ, A^2 is simply proportional to E/ρ_0.

So $E \propto \rho_0 A^2$. Combined with the definition in Eq. 1.29 of A, this is just the scaling law of Eq. 1.23. However, the constant of proportionality is now known, in terms of integrals over the functions f, ψ, and ϕ. Under these conditions we can write another form of Eq. 1.44 as follows:

$$E = 2\pi\int_0^1 \psi\phi^2\eta^2 d\eta + \frac{4\pi}{\gamma(\gamma-1)}\int_0^1 f\eta^2 d\eta \tag{1.44}$$

With this result and the experimental comparisons, Taylor's paper provided the first important work on the theory of intense explosions. While he missed the analytic solution, Taylor produced a hugely valuable numerical solution. The utility of the method he used remains; perturbations of this solution and similar scaling arguments are ubiquitous in astrophysics and other areas. Numerical solution for $\gamma = 1.4$ under assumption of an adiabatic condition for surrounding atmosphere of high intense point blast is given by Taylor [20] in his famous published paper.

The above dimensional analysis for this example was given based on "The Formation of a Blast Wave by a Very Intense Explosion" going outward. As we saw Taylor has done a fantastic job to formulate such problem and derive an excellent conclusion that is close to Trinity nuclear explosion observations. However, what will happen if the blast wave travels inward? This is subject of our next section, and it is known as Karl Gottfried Guderley [26] problem since he was the first physicist and mathematician from Germany that looked at this issue in 1942. He was able to solve this problem in one-dimensional coordinate, in spherical and cylindrical format using dimensional analysis method in contrast to Taylor's problem. Later on in 1954, Butler [27] published his report dealing with converging spherical and cylindrical shock.

1.11 Energy in a High Intense Implosion

The focusing of spherical and cylindrical shock waves has been for various engineering and scientific applications. It is an effective and economic means of producing high temperature and pressure gas at the implosion center as focal point. For instance, converging shock-wave process has been successfully adopted in the production of high temperature and density plasma as part of *laser-driven fusion program* [28–30] for imploding pellet of hydrogen isotopes such as deuterium and tritium. Or for that matter, in man-made explosion of the *Physical Principles of Thermonuclear Explosive Devices* [31] or in nature such as supernovae [32] or other application such as synthetic diamonds from graphite carbide and neutrons [33]. Moreover, they are used in research related to particles lunched at hypersonic velocity. They are also associated with research related to substance behavior under severe conditions in a high-energy medium [34].

Analytically, Guderley [26] was the first to present a comprehensive investigation involving cylindrical and spherical shock-wave propagation in air, and he obtained a *similarity* solution. In his solution shock strength was found to be proportional to R^n with R being the distance of shock from the center of implosion and n a constant that depends on adiabatic index γ which is specific heat ratio which is shown as before ($\gamma = C_P/C_V$). This model clearly implies that theoretically a converging shock wave can increase in strength indefinitely as the radius R approaches zero. In practice, at the center of implosion, temperature and pressure can attain very high but finite values due to experimental limitation.

Following Guderley's theory, other scientists such as Butler [27] and Stanyukovich [33] focused their effort on the development of similarity solutions for the converging process.

Guderley has analyzed the flow behind a converging spherical or cylindrical shock. His treatment of the incoming shock and the flow immediately behind it is complete, but less attention seems to have been paid to the reflected shock and the associated region of disturbance. Butler's paper presents the physical assumptions underlying Guderley's analysis of the incoming shock which are clarified, and the reflected shock is treated.

Ashraf [35] is considering imploding spherical and cylindrical shocks near the center (axis) of implosion when the flow assumes a self-similar character. The shock becomes stronger as it converges toward the center (axis), and there is high temperature behind the shock leading to intense exchange of heat by radiation or condition. His assumption is that the flow behind the shock is not adiabatic but is approximately isothermal, and the time-dependent temperature behind the shock goes on changing as the shock propagates, and this temperature is different from that ahead of the shock. The flow behind the shock is likely to have nearly uniform spatial distribution and that is why the temperature gradient is considered to be zero. This type of flow is known as "homo-thermal flows" and has been dealt by scientists. Except for the idealized intense heat exchange behind the shock, the problem is the same as has been discussed by Guderley.

All of them obtained similarity solutions by reducing the problem to nonlinear first-order differential equations. The similarity exponent δ of the shock trajectory $R \propto (-t)^{\delta}$, where t is the time taken by the shock to cover the distance R to reach the origin, cannot be evaluated from dimensional considerations as occurs in the Taylor's [20] explosion problem. The time t taken to be negative before the shock converges to the center (axis) of symmetry and $t = 0$ is instant at which the shock converges to the center (axis).

The shock position is assumed to be given by Eq. 1.45:

$$R = A(t)^{\alpha} \quad \xi = \frac{r}{R} = \frac{r}{A(-t)^{\alpha}} \tag{1.45}$$

where R is the radial distance of the shock from center (axis) and A along with α is a positive constant. The interval for variables that are involved with solution that we are seeking includes all of space up to infinity, so that the intervals for the variables are

$$-\infty < t \leq 0 \quad R \leq r < \infty \quad 1 \leq \xi < \infty$$

Note that the self-similar solution holds only in a region with radius of the order of R, and at large distance it is connected with the solution of the complete nonself-similar problem in some manner [25].

At the shock front $\xi = 1$, and the front velocity is directed toward the center and is negative, with $\dot{R} = \alpha R/t = -\alpha R/|t| < 0$.

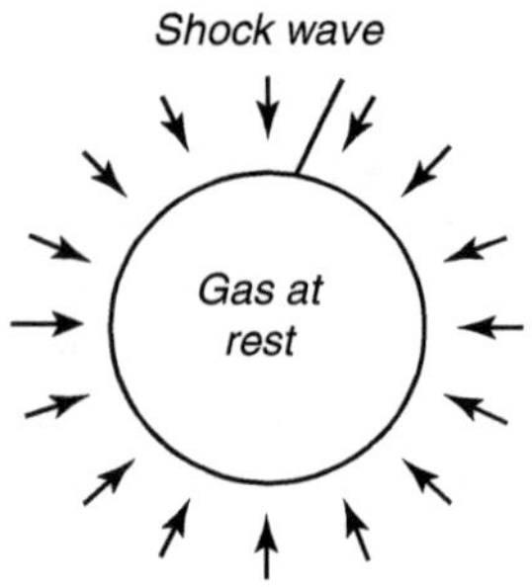

Fig. 1.15 Diagram of motion of implosion point in front of the core adiabatic compression

The basic equations governing the one-dimensional flow in terms of Lagrangian coordinate η and time t, having temperature as function of time only, are (Eq. 1.46) (Fig. 1.15)

$$\begin{aligned}\frac{\partial r}{\partial \eta} &= \frac{1}{\rho r^{j-1}} \\ \frac{\partial u}{\partial t} &= -r^{j-1}\frac{\partial p}{\partial \eta} \\ \frac{\partial T}{\partial r} &= 0\end{aligned} \tag{1.46}$$

where r is the distance of the particle from the center or axis of symmetry and u, ρ, p, and T are the particle velocity, density, pressure, and temperature behind the shock wave, respectively. η is the Lagrangian coordinate defined by Eq. 1.47:

$$d\eta = \rho_0 r_0^{j-1} dr_0 \tag{1.47}$$

Note
The definition and concept of both Lagrangian and Eulerian are described in Chap. 2, Sect. 1.2 of this book.

where ρ_0 is the ambient density, r_0 is the value of r at the initial instant of time, and $j = 2, 3$ for cylindrical and spherical symmetry, respectively.

The continuity equation of (first set) of Eq. 1.46 may be expressed in terms of particle velocity u as Eq. 1.48:

$$\frac{\partial u}{\partial \eta} = -\frac{1}{\rho^2 r^{j-1}}\left[\frac{\partial \rho}{\partial t} + (j-1)\frac{\rho u}{r}\right] \tag{1.48}$$

Table 1.7 Values of α obtained from the analytic expression

	Cylindrical	Spherical
	$j=2$	$j=3$
6/5	0.7767	0.6349
7/5	0.7285	0.5730
5/3	0.6978	0.5359

For strong shock we can now establish the boundary conditions at the shock assuming that the radiative flux across the optically thin shock layer is continuous so that the classical shock condition holds [35], and they are written as follows:

$$\begin{cases} u_s = (1-\beta)\dot{R} \\ \rho_s = \rho_0/\beta \\ p_s - (1-\beta)\rho_0\dot{R}^2 \end{cases}$$

where ρ_0 is the ambient density, $\dot{R}\cdot$ the shock velocity, and β the density ratio across a strong shock and is equal to $\frac{\gamma-1}{\gamma+1}$.

Ashraf [35] is carrying on a self-similar solution in this case when he introduces a similarity variable $\mu = \frac{\eta}{\eta_s}$. In his analysis, he seeks a closed form solution via an approximate analytic approach and shows this solution up to second-order terms in β. He also demonstrates for the zero-order approximation the particle velocity which is same as the shock velocity, while the density and pressure are linear function of μ, which implies that the first- and second-order terms from his established equations contribute more significantly to velocity than to the density and pressure. His solution also describes the Eulerian distance r and Eulerian similarity variable ξ as function of Lagrangian similarity variable μ, which also will indicate that zeroth approximation of the Eulerian distance is same as the shock distance.

Analysis of differential equations of gas dynamics associated with this problem allows the similarity exponent α to be obtained by solving these differential equations either *analytically* or numerically.

His finding for values of similarity exponent α for the adiabatic flow both analytically and numerically is shown in the Tables 1.7 and 1.8 as follows:

He finds that the value of similarity exponent α is smaller for the homo-thermal flows that the adiabatic flows so that the shock velocity $\dot{R} \propto R^{-(1-\alpha)/\alpha}$ is larger in the former case than in the latter as the shock approaches the center or axis. The shock velocity and hence pressure tend to infinity as $t \longrightarrow 0$ being the instant of shock implosion.

Zel'dovich and Raizer [25] also show as the shock converges, energy becomes concentrated near the shock front as the temperature and pressure there increase without limit, but the dimensions of the self-similar region decrease with time. They consider a self-similar solution within sphere whose radius decreases in proportion to the radius of the front R. The effective boundary of this similar region is then considered to be at some constant value $r/R = \xi = \xi_1$.

Table 1.8 Values of α obtained from numerical integration

	Cylindrical	Spherical
6/5	0.7963	0.6553
7/5	0.7530	0.5965
5/3	0.7218	0.5589

They present an equation for energy contained in spherical implosion situation with the variable radius $r_1 = \xi_1 R$ as follows (Eq. 1.49):

$$\begin{aligned} E_{S_{\text{implosion}}} &= \int_R^{r_1} 4\pi r^2 dr \rho \left(\frac{1}{\gamma - 1} \frac{p}{\rho} + \frac{u^2}{2} \right) \\ &= 4\pi R^3 \rho_0 \dot{R}^2 \int_1^{\xi_1} g \left(\frac{1}{\gamma - 1} \frac{\pi}{g} + \frac{v^2}{2} \right) \xi^2 d\xi \end{aligned} \tag{1.49}$$

The integral with respect to ξ from 1 to ξ_1 is a constant, so that the energy $E_{S_{\text{implosion}}} \sim R^3 \dot{R}^2 \sim R^{5-(2/\alpha)}$. The exponent of R is positive for all real values of the specific heat ratio (adiabatic index) γ. For example, for $\gamma = 7/5$, the similar exponent $\alpha = 0.717$, which is close to what is shown by Ashraf [35] calculation in Tables 1.7 and 1.8:

$$E_{S_{\text{implosiion}}} \sim R^{2.21} \to 0 \quad \text{as} \quad R \to 0 \tag{1.50}$$

With integration with respect to ξ extended to infinity ($\xi_1 = \infty$) the integral diverges, thus the total energy in all space is infinite within the framework of the self-similar solution.

In summary, in order to find the value of α, numerically do a trial and error analysis, where a value of α is assumed and related differential equation is integrated numerically from the initial point A ($\xi = 1$), and the behavior of the integral curve is determined. In our value, the case for $\gamma = 7/5$, and the limiting density is (behind the shock front $\rho_1 = 6\rho_0$) about $\rho_{\text{limit}} = 21.6\rho_0$. The density at large distance from the front $r \to \infty$ before the instant of collapse at the center (or axis) is also $\rho = 21.6\rho_0$, since for $R \neq 0$ and $r \to \infty$, $\xi = r/R \to \infty$ and $\rho/\rho_0 = G(\xi) \to G(\infty)$. Under these conditions, the collapsed energy concentrated at the center of sphere is given by [25]

$$\int_0^r 4\pi r^2 dr \rho \left(\frac{1}{\gamma - 1} \frac{p}{\rho} + \frac{u^2}{2} \right) \sim r^{5-2/\alpha} \tag{1.51}$$

just as $E_{S_{\text{implosion}}} \sim R^{5-2/\alpha}$ which is seen in above.

1.12 Similarity and Estimating

The notion of similarity is familiar from geometry. Two triangles are said to be similar if all of their angles are equal, even if the sides of the two triangles are of different lengths. The two triangles have the same shape; the larger one is simply a scaled-up version of the smaller one. This notion can be generalized to include physical phenomena. This is important when *modeling* physical phenomena, for instance, testing a prototype of a plane with a scale model in a wind tunnel. The design of the model is dictated by dimensional analysis. Similarity is an extension of geometrical similarity. By definition, two systems are similar if their corresponding variables are proportional at corresponding locations and times. The famous of all and familiar similarity that one can even buy in today's market is Russian nested dolls.

A *matryoshka doll* or a *Russian nested doll* (often incorrectly referred to as a *babushka doll*—babushka means "grandmother" in Russian) is a set of dolls of decreasing sizes placed one inside the other. "Matryoshka" (Матрёшка) is a derivative of the Russian female first name "Matryona," which was a very popular name among peasants in old Russia. The name "Matryona" in turn is related to the Latin root "mater" and means "mother," so the name is closely connected with motherhood, and in turn the doll has come to symbolize fertility.

A set of matryoshkas consists of a wooden figure, which can be pulled apart to reveal another figure of the same sort inside. It has, in turn, another figure inside, and so on. The number of nested figures is usually five or more. The shape is mostly cylindrical, rounded at the top for the head, and tapered toward the bottom, but little else; the dolls have no hands (except those that are painted). Traditionally the outer layer is a woman, dressed in a sarafan. Inside, it contains other figures that may be of both genders, usually ending in a baby that does not open. The artistry is in the painting of each doll, which can be extremely elaborate. See the figure below.

Return to the mathematical statement of the Π theorem, Eq. 1.12. We can identify the following dimensionless parameters:

$$\Pi = \frac{a_n}{a_1^p \ldots a_k^r} \qquad \Pi_1 = \frac{a_{k+1}}{a_1^{p_{k+1}} \ldots a_k^{r_{k+1}}} \tag{1.52}$$

and so on, such that Eq. 1.12 can be written as

$$\Pi = \Phi(\Pi_1, \ldots, \Pi_{n-k})$$

The parameters $(\Pi, \Pi_1, \ldots, \Pi_{n-k})$ are known as *similarity parameters*. Now if two physical phenomena are similar, they will be described by the same function Φ. Denote the similarity parameters of the model and the prototype by the superscripts m and p, respectively. Then if the two are similar, their similarity parameters are equal:

$$\Pi_1^{(p)} = \Pi_1^{(m)}, \ldots, \Pi_{n-k}^{(p)} = \Pi_{n-k}^{(m)} \tag{1.53}$$

So that

$$\Pi^{(p)} = \Pi^{(m)} \tag{1.54}$$

Therefore, in order to have an accurate physical model of a prototype, we must first identify all of the similarity parameters and then insure that they are equal for the model and the prototype.

Finally, we come to estimating. In this course we will often make order-of-*magnitude estimates*, where we try to obtain an estimate to within a factor of ten (sometimes better, sometimes worse). This means that we often drop factors of two, although we should exercise some caution in doing this. Estimating in this fashion is often aided by first doing some dimensional analysis. Once we know how the governed parameter (which we are trying to estimate) scales with other quantities, we can often use our own personal experience as a guide in making the estimate. Similarity is one of the most fundamental concepts, both in physics and mathematics. This first aspect, geometrical similitude, is the best known, the best understood, and another one, more abstract, deals with the physical similitude. Since all systems must obey the same physical laws, in addition to the geometrical scaling factors, relations between different physical quantities must be fulfilled in order to make two systems similar. Again, Russian nested dolls (Fig. 1.16) are a very good example of such similarity. We recognize how central will be these ideas in the theory of modeling. Such reduced models play a central role in shipbuilding, aeronautical engineering, oceanography, etc. In engineering quite often, many different phenomena, belonging to different branches of science, take place simultaneously, and conflicts are possible. However, other aspects of similarity can be

Fig. 1.16 Russian nested dolls

found in the logic of a machine or in an algorithm. Under its geometrical and logical aspects, similarity and self-similarity appear as rather regular, easy to distinguish patterns. Nature has more fantasy, and in some cases, it likes to add some randomness. Self-similarity, in that case, is more difficult to distinguish but is still there.

1.13 Self-Similarity

Now that we are here, the question is what is self-similarity? Simply we can use the answer that is given in Wikipedia, and it seems a good description of it.

"In mathematics, a self-similar object is exactly or approximately similar to a part of itself (i.e. the whole has the same shape as one or more of the parts). Many objects in the real world, such as coastlines, are statistically self-similar: parts of them show the same statistical properties at many scales. Self-similarity is a typical property of fractals.

Scale invariance is an exact form of self-similarity where at any magnification there is a smaller piece of the object that is similar to the whole. For instance, a side of the Koch snowflake is both symmetrical and scale-invariant; it can be continually magnified $3\times$ without changing shape."

So in a simple form, self-similarity means that the form of the solutions is scaled invariant (*temporally* and *spatially*). Dealing with astrophysical hydrodynamics problem, mostly supernova and strong shock phenomenon, we encounter "simila" or "self-similar" solutions and using "similarity methods."

Self-similarity means that a structure, or a process, and a part of it appear to be the same when compared. A self-similar structure is infinite and it is not differentiable in any point.

In physics and mathematics, scale invariance is a feature of objects or laws that do not change if length scales (or energy scales) are multiplied by a common factor. The technical term for this transformation is a dilatation (also known as dilation), and the dilatations can also form part of a larger conformal symmetry.

- In mathematics, scale invariance usually refers to an invariance of individual functions or curves. A closely related concept is self-similarity, where a function or curve is invariant under a discrete subset of the dilatations. It is also possible for the probability distributions of random processes to display this kind of scale invariance or self-similarity.
- In classical field theory, scale invariance most commonly applies to the invariance of a whole theory under dilatations. Such theories typically describe classical physical processes with no characteristic length scale.
- In quantum field theory, scale invariance has an interpretation in terms of particle physics. In a scale-invariant theory, the strength of particle interactions does not depend on the energy of the particles involved.
- In statistical mechanics, scale invariance is a feature of phase transitions. The key observation is that near a phase transition or critical point, fluctuations occur at all length scales, and thus one should look for an explicitly scale-invariant

theory to describe the phenomena. Such theories are scale-invariant statistical field theories and are formally very similar to scale-invariant quantum field theories.

- Universality is the observation that widely different microscopic systems can display the same behavior at a phase transition. Thus, phase transitions in many different systems may be described by the same underlying scale-invariant theory.
- In general, dimensionless quantities are scale invariant. The analogous concept in statistics is standardized moments, which are scale-invariant statistics of a variable, while the unstandardized moments are not.

The self-similarity can be grouped to the following general categories as follows [4]:

1. *Approximate self-similarity*: means that the object does not display perfect self-similarity. For example, a coastline is a self-similar object, a natural fractal, but it does not have perfect self-similarity. A map of a coastline consists of bays and headlands, but when magnified, the coastline is not identical, but statistically the average proportions of bays and headlands remain the same no matter the scale [6]. It is not only natural fractals that display approximate self-similarity; the Mandelbrot set is another example. Identical pictures do not appear straight away, but when magnified, smaller examples will appear at all levels of magnification [5, 6].
2. *Statistical self-similarity:* means that the degree of complexity repeats at different scales instead of geometric patterns. Many natural objects are statistically self-similar, whereas artificial fractals are geometrically self-similar [36].
3. *Geometrical similarity:* is a property of the space-time metric, whereas physical similarity is a property of the matter fields. The classical shapes of geometry do not have this property; a circle if on a large enough scale will look like a straight line. This is why people believed that the world was a flat pancake; the earth just looks that way to humans [6, 36, 37].

One well-known example of self-similarity and scale invariance is fractals, patterns that form of smaller objects that look the same when magnified. Many natural forms, such as coastlines, fault and joint systems, folds, layering, topographic features, turbulent water flows, drainage patterns, clouds, trees, leaves, bacteria cultures [36], blood vessels, broccoli, roots, lungs and even universe, etc., look alike on many scales [37].

Let us see what experts such as Barenblatt [10] are saying about self-similarity and how they describe it.

Although, in general, self-similarity may be expressed in several different ways, it is often manifested mathematically as a power function $y = ax^{\beta}$, which obeys the homogeneity relation $y(\lambda x) = \lambda^{\beta} y(x)$, where λ is a (positive) scale factor and β is a scaling exponent. Functions that satisfy this relation are said to be scaling functions, while processes or objects that are described by such functions are said to exhibit scaling behavior. With this, the terms scaling, scale invariance, and self-similarity are often used as interchangeable terms. There are numerous examples of power relationships between geological variables (Turcotte) [2], though the ranges of reported scaling behavior are often less than one order of magnitude. Actually, this is not surprising, as scaling behavior in nature is always limited between internal (small) and external (large) scales introduced by the driving mechanisms or by structural properties. A good example may be found in turbulence where classical Kolmogorov's scaling (Monin and Yaglom [25]; Frisch [19]) is constrained by viscosity at small scales and by the flow size at large scales. With limited data, such constraints introduce unavoidable uncertainties in the identification of true scaling behavior or scaling regions (Avnir and others [26]).

Self-similarity is a special condition of a single system. A system is said to be self-similar if there exists a separable variable of the principal equations and initial and boundary conditions of the system. The separable variable is called a similarity variable. Similarity variables are valuable in the solution of special partial differential equations with special initial and boundary conditions. Solutions of the diffusion equation and the *Prandtl* boundary-layer equations are classical examples of the application of similarity variables [38].

Self-similar solutions provide some of the greatest simplifications to one-dimensional flows. Self-similarity allows the reduction of the partial differential equations, which contain two independent variables (space and time), into a set of ordinary differential equations (ODEs), where the single independent variable is a combination of space and time. The ODEs are then relatively easy to solve numerically or even analytically in some cases. They describe the asymptotic behavior of one-dimensional flow, in a variety of circumstances. Typically, they are far away from the initial conditions and provided that the boundary conditions contain no spatial scale.

Some exceptions apply. For example, self-similarity can prevail in exponential density gradient in planar geometry.

Whether or not a system is self-similar is not obvious, and the discovery of similarity variables may be a tedious process. Two approaches may be followed. The first one starts with the initial and boundary conditions. The second one starts with the principal equations. The first approach is simpler, if it is known, or assumed similarity variables apply. The second approach may reveal a more general class of separable variables, which may or may not satisfy specified initial and boundary conditions.

A typical initial or boundary condition of self-similar systems is that

$$f(a,y) = f(x,b) \tag{1.55}$$

where x and y may be either coordinates or time. Eq. 1.55 may be satisfied, in some cases, by a similarity variable of the form

$$\zeta = x^m y^n \tag{1.56}$$

Besides geometrical similarity, the first one to recognize a coherent structure in a physical phenomenon was Fourier with his study of the heat propagation. Then, mostly the fluid dynamists of the late nineteenth to beginning of the twentieth century recognize the idea of physical similarity between different experiments and the possibility of comparing their results after the introduction of properly chosen dimensionless quantities. From these works, it emerges the concept of dimensional analysis with Π theorem. At the same time, reduced models are used in engineering. Invariance under similarity transformation and/or under time translation is quite common properties in the equations modeling the physical world. They allow partial or total integration and lead to much simpler equations, which eventually can be solved numerically with a much-reduced numerical effort (decrease in the dimension of the phase space or the parameters space or decrease in the number of independent variables). Nevertheless similarity transformations and time translations put constraints on the initial conditions which can be treated although they often point out these initial conditions or the critical parameters for which the nature of the solution changes. Embedding these concepts in the physical frame of rescaling can permit to precise the nature of these self-similar solutions (SSS) and give information on their possible asymptotic nature. In that case the knowledge of the physicists complements nicely the more rigorous mathematical treatment.

The following example that is from Skoglund [38] book involves the classical one-dimensional transient diffusion equation. A separation variable is determined before considering the initial and boundary conditions.

Example 1.9: Impulsive Flat Plate in a Fluid At time $t < 0$, an infinite flat plate and the incompressible viscous fluid above it are at rest. At time $t > 0$, the constant velocity of the plate in the x direction is u_0.

Specification: The initial and boundary conditions are

$$u(y,0) = u(\infty,t) = 0 \quad u(0,t) = u_0 \tag{1}$$

where u = fluid velocity in the x direction
y = distance normal to the plate
t = time
u_0 = constant velocity of the plate for $t > 0$

Neglecting convective terms of the Navier–Stokes equation,

$$\frac{\partial u}{\partial t} = \nu \frac{\partial^2 u}{\partial y^2} \tag{2}$$

where ν = kinematic viscosity

Problem: Derive the similarity variables of Eqs. 1 and 2 of Example 1.9.

Calculation: Nondimensionally

$$f''(0) \approx 0.332 \tag{3}$$

$$n(x,0) = n(\infty,\tau) = 0 \quad n(0,\tau) = 1 \tag{4}$$

where

$$n = \frac{u}{u_0}$$

$$N_0 = \frac{vt_0}{x_0^2} = 1$$

$$\chi = \frac{y}{x_0}$$

$$\tau = \frac{t}{t_0}$$

Assuming that $n(\chi, \tau) = n(\zeta)$

$$\frac{\partial n}{\partial \tau} = n'\zeta \quad \frac{\partial^2 n}{\partial \chi^2} = n'\zeta_{xx} + n''(\zeta_x)^2 \tag{5}$$

where

$$n' = \frac{dn}{d\zeta} \quad n'' = \frac{d^2 n}{d\zeta^2}$$

$$\zeta_r = \frac{\partial \zeta}{\partial \tau}$$

$$\zeta_r = \frac{\partial \zeta}{\partial \chi} \quad \zeta_{xx} = \frac{\partial^2 \zeta}{\partial \chi^2}$$

By combining Eqs. 3 and 5 we have

$$n'(\zeta_\tau - \zeta_{xx}) = n''(\zeta_x)^2 \tag{6}$$

As suggested by Equation-by-Equation 1.49, let $\zeta = \tau^m \chi^n$. To simplify Eq. 6, let $n = 1$, so $\zeta = \chi\tau^m$, and

$$\zeta_\tau = m\chi\tau^{m-1} = m\zeta\tau^{-1} \quad \zeta_x = \tau^m \quad \zeta_{xx} = 0 \tag{7}$$

By combining Eqs. 6 and 7, we have

$$n'm\zeta = n''\tau^{2m+1} \tag{8}$$

The following ordinary differential equation is obtained by letting $2m + 1 = 0$ or $m = -1/2$

$$n'' + \frac{1}{2}\zeta n' = 0 \tag{9}$$

From Eq. 4, we see that

$$n(0) = 1 \quad n(\infty) = 0 \tag{10}$$

A solution of Eqs. 9 and 10 is

$$n(\chi, \tau) = 1 - \mathrm{erf}(\zeta/2) \tag{11}$$

where

$$\mathrm{erf}(x) = \frac{2}{\sqrt{\pi}}\int_0^x e^{-t^2 dt}$$
$$\mathrm{erf}(0) = 0$$
$$\mathrm{erf}(\infty) = 0$$

Discussion: In an unusual problem, some trial and error might be necessary in selecting possible similarity variables, but the above procedure would reduce the number of possible alternatives. In this case, the number of similarity variables was less than the original number of variables. As indicated by the next example, that is not always true. In addition, we have discussed more details in Chap. 4, about solving nonlinear partial differential equation via self-similarity by using *Lie group theory* to reduce them to an ordinary differential equation form.

Now next example involves Blasius boundary layer. The boundary conditions of it are utilized first, because the separation of variables is complicated.

Blasius Boundary Layer

A Blasius boundary layer, in physics and fluid mechanics, describes the steady two-dimensional boundary layer that forms on a semi-infinite plate which is held parallel to a constant unidirectional flow U.

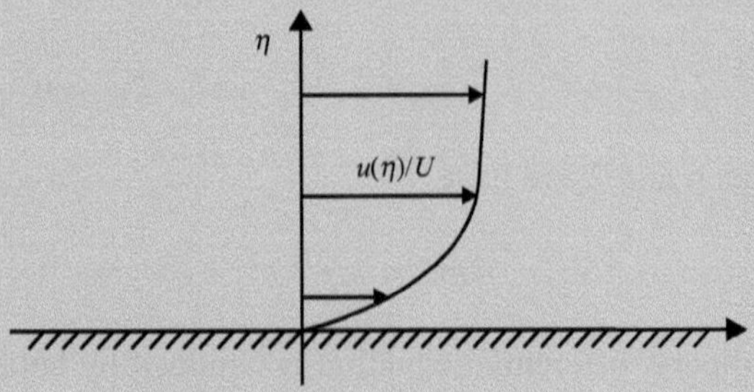

A schematic diagram of the Blasius flow profile. The stream-wise velocity component $u(\eta)/U(x)$ is shown, as a function of the stretched coordinate η.

(continued)

(continued)

The solution to the Navier–Stokes equation for this flow begins with an order-of-magnitude analysis to determine what terms are important. Within the boundary layer, the usual balance between viscosity and convective inertia is struck, resulting in the scaling argument:

$$\frac{U^2}{L} \approx \nu \frac{U}{\delta^2},$$

where δ is the boundary-layer thickness and ν is the kinematic viscosity.

However the semi-infinite plate has no natural length scale L and so the steady, incompressible, two-dimensional boundary-layer equations for continuity and momentum are

$$\frac{\partial u}{\partial x} + \frac{\partial v}{\partial y} = 0$$

$$u\frac{\partial u}{\partial x} + v\frac{\partial u}{\partial y} = \nu \frac{\partial^2 u}{\partial y^2}$$

(note that the x independence of U has been accounted for in the boundary-layer equations) admit a similarity solution. In the system of partial differential equations written above, it is assumed that a fixed solid body wall is parallel to the x direction, whereas the y direction is normal with respect to the fixed wall. u and v denote here the x and y components of the fluid velocity vector. Furthermore, from the scaling argument, it is apparent that the boundary layer grows with the downstream coordinate x, e.g.,

$$\delta(x) \approx \left(\frac{\nu x}{U}\right)^{1/2}$$

This suggests adopting the similarity variable

$$\eta = \frac{y}{\delta(x)} = y\left(\frac{U}{\nu x}\right)$$

and writing

$$u = Uf(\eta)$$

It proves convenient to work with the stream function Ψ, in which case

$$\psi = (\nu U x)^{1/2} f(\eta)$$

(continued)

(continued)

and on differentiating, to find the velocities, and substituting into the boundary-layer equation, we obtain the Blasius equation:

$$f''' + \frac{1}{2}ff'' = 0$$

subject to $f = f' = 0$ on $\eta = 0$ and $f' \to 1$ as $\eta \to \infty$. This nonlinear ordinary differential equation must be solved numerically, with the shooting method proving an effective choice. The shear stress on the plate

$$\tau_{xy} = \frac{f''(0)\rho U^2 \sqrt{\nu}}{\sqrt{Ux}}$$

can then be computed. The numerical solution gives $f''(0) \approx 0.332$

Example 1.10: Blasius Boundary Layer A Blasius boundary layer is steady, incompressible laminar boundary layer on a smooth flat plate that satisfies Prandtl boundary-layer equations. The following example is only an outline of the mathematics of discovering the similarity variables of a Blasius boundary layer. Details of boundary-layer theory may be found in a book by Schlichting [39] and what we reflect here in general from Wikipedia.

Specifications: To simplify the symbolism, normalized conditions are represented by the usual dimensional symbols. The initial and boundary conditions are

$$u(0, y) = u(x, \infty) = 1 \tag{1}$$

$$u(x, 0) = 0 \tag{2}$$

$$\upsilon(0, y) = \upsilon(x, 0) = 0 \tag{3}$$

where $u(x, y)$ = the nondimensional x component of velocity parallel to the plate

$\upsilon(x, y)$ = the nondimensional y component of velocity perpendicular to the plate

Problem: Derive the similarity variables.

Principle: From Schlichting [39], a Prandtl boundary-layer equation is

$$n_y n_{xy} - n_x n_{yy} = n_{yyy} \tag{4}$$

where $n_y = \partial n/\partial y = u$

$n_x = \partial n/\partial x = -\upsilon$

$n(x, y)$ = a nondimensional stream function

Calculations: To separate the variables, let

$$n(x,y) = f(\chi)g(\zeta) \tag{5}$$

where $\chi(x, y)$ and $\zeta(x, y)$ are transformed coordinates. From Eq. 1 in this example, we have

$$n_y(0,y) = n_y(x,\infty) = 1 \tag{6}$$

By combining Eqs. 5 and 6 for $(0, y)$ or (x, ∞)

$$f'g\chi_y + fg'\zeta_y = 1 \tag{7}$$

where

$$f' = \frac{df}{d\chi}$$
$$g' = \frac{dg}{d\zeta}$$
$$\chi_y = \frac{\partial \chi}{\partial y}$$
$$\zeta_y = \frac{\partial \zeta}{\partial y}$$

To simplify, let $\chi_y = 0$, so that $\chi = \chi(x)$, and $f(\chi) = f(x)$. From Eq. 7, we have

$$n_y(0,y) = n_y(x,\infty) = f(x)g'(\zeta)\zeta_y = 0 \tag{8}$$

Comparison of Eqs. 8 and 1.49 suggests that $\zeta = y^n x^m$. To simplify Eq. 7, let $f(x) = (\zeta_y)^{-1} = x^{-m}$, so that $\zeta = yx^m$. For $m < 0$

$$\zeta(0,y) = \zeta(x,\infty) = \infty \tag{9}$$

Thus, the boundary conditions of Eq. 6 are satisfied by $f(\chi) = x^{-m}$, $\zeta = yx^m$, and $g'(\infty) = 1$.

The variables of x and ζ are separated in Eq. 4 if $m = -1/2$ which checks the assumption of Eq. 9 of $m < 0$. The ordinary differential equation, which replaces Eq. 4, is

$$gg'' + 2g''' = 0 \tag{10}$$

To satisfy the boundary conditions of Eqs. 2 and 3

$$n_y(x, 0) = fg'\zeta_y = g'(0) = 0 \tag{11}$$

$$n_x(x, 0) = f'g + fg'\zeta_x = f'g(0) = 0 \tag{12}$$

In summary, the boundary conditions of Eq. 10 are

$$g'(0) = g(0) = 0 \quad \text{and} \quad g'(\infty) = 1 \tag{13}$$

Equations 10 and 13 have been solved approximately, and the results are in good agreement with measurements. In this case, $f(x) = \sqrt{x}$ is simple, and the results may be expressed in a single curve of $n/\sqrt{x} = g(\zeta)$ versus $\zeta = y/\sqrt{x}$.

1.14 General Results of Similarity

If the general requirements of similarity are satisfied, the solution of the nondimensional equations will be the same for a prototype and its model. Therefore, the result of similarity is that corresponding, dependent, nondimensional variables are equal at corresponding points. Mathematically, the result of similarity is that

$$n_{\mathrm{D}}(\chi_i, \tau)_p = n_{\mathrm{D}}(\chi_i, \tau)_m \tag{1.57}$$

where n_{D} is a dependent nondimensional variable.

1.14.1 Principles of Similarity

By combining the requirements and results of similarity, the principles of similarity are

$$N_{\mathrm{om}} = N_{\mathrm{op}} \quad \text{and} \quad n_u(\chi, \tau)_m = n_u(\chi, \tau)_p \tag{1.58}$$

where N_{om} = model reference similarity number
N_{op} = prototype reference similarity number
u = dimensional variable
n = nondimensional variable
χ_i = nondimensional coordinate
τ = nondimensional time

$n_u(\chi_i, \tau)_m = n_u(\chi_i, \tau)_p$ includes geometrical requirements.

To derive a corollary to the principle of similarity, consider a local similarity number:

$$N(\chi_i, \tau) = u_1^{\alpha} u_2^{\beta} \ldots u_n^{\eta} \tag{1.59}$$

By combining Eq. 1.59 with Eq. 1.58, we get Eq. 1.60 as follows:

$$\left(n_{u1}^{\alpha} n_{u2}^{\beta} \ldots n_{un}^{\eta}\right)_p = \left(n_{u1}^{\alpha} n_{u2}^{\beta} \ldots n_{un}^{\eta}\right)_m \tag{1.60}$$

Since $n_u = u/u_0$

$$\frac{\left(u_1^{\alpha} u_2^{\beta} \ldots u_n^{\eta}\right)_p}{\left(n_{10}^{\alpha} n_{20}^{\beta} \ldots n_{n0}^{\eta}\right)_p} = \frac{\left(n_1^{\alpha} n_2^{\beta} \ldots n_n^{\eta}\right)_m}{\left(n_{10}^{\alpha} n_{20}^{\beta} \ldots n_{20}^{\eta}\right)_m} \tag{1.61}$$

However, since $N_{\text{op}} = N_{\text{om}}$, the denominators are equal, and the numerators

$$N(\chi, \tau)_p = N(\chi, \tau)_m \tag{1.62}$$

Therefore, a corollary to the principle of similarity is that corresponding local similarity numbers are equal at corresponding points of similar systems. Local similarity numbers do not involve reference variables and are useful in some applications.

1.15 Scaling Argument

Scaling in this case is a technique of relating the prototype and model variables at corresponding points. From Eq. 1.58, a scaling factor is

$$k_u = \frac{u(\chi, \tau)_p}{u(\chi, \tau)_m} = \frac{u_{\text{op}}}{u_{\text{om}}} = \text{constant} \tag{1.63}$$

Therefore, the ratio of the corresponding dimensional variables of a prototype and model at corresponding points is constant. Scaling factors are commonly used in the design and interpretation of models, but they obscure the basis of similarity.

1.16 Self-Similar Solutions of the First and Second Kind

We have learned so far that two geometrical objects are called similar if they both have the same shape. The second object may be obtained from the first by the result of a uniform scaling (enlarging or shrinking). Also from all demonstrations and example presented in the above section, we have established for certain engineering and physics problems that we cannot find a close analytical solution; therefore, dimensional analysis is a good tool to use. Meanwhile we observed how extensions of this tool such as scaling and similarity, and also self-similarity have great influence to establish a solution for these nonlinear problems. Now we need to take one step beyond where we are by defining different types of self-similarities in particular when we are dealing with typical gas dynamic and fluid mechanics where nonlinear ordinary or partial differential equations present themselves. For example, in gas dynamics, two types of self-similar process, termed as self-similar motions of the first kind and the second kind, have been considered by scientist such as Zel'dovich and Raizer [25] and G. I. Barenblatt [40]. Taylor's explosion problem and one-dimensional centered rarefaction waves are typical scenarios of the flows of this kind, while emergence of strong shock near the surface of star, Sakurai [41], Sachdev and Ashraf [42], and converging cylindrical and spherical shocks Guderley [26], are examples of the flows of the second kind.

Zel'dovich and Raizer [25] are suggesting that the solutions of the first type possess the property that the similarity exponent α and the exponent of t and R in all scales are determined either by dimensional considerations or from the conservation laws [25]. They also describe under these circumstances the exponents are simple rational fractions with integral numerators and denominators. They explain the problem of this type always contains two parameters with independent dimensions, which means there is a type of self-similar solution in which the exponents are determined by the boundary conditions and may be set arbitrary within certain limits. Although the exponents in such solutions are not simple rational fractions in general, the solutions are to be considered as the first type, because the two independent parameters exist and the exponents are determinable in advance. These parameters are used to construct a parameter a whose dimension contains the primary type units, which is mass and the other two parameters of length and time, and it is designated by A. With the latter parameter, A, it is possible to construct a dimensionless combination, the similarity variable $\xi = r/At^{\alpha}$. The dimensions of the parameter A are given in terms of length and time as $LT^{-\alpha}$ which are determined by the similarity exponent α. Examples of these types were well presented above (i.e., Taylor explosion problem) [20, 21].

Taylor [20], von Neumann [43], and Sedov [44], known as the Sedov–Taylor solution, describe an explosion in which a strong shock wave propagates into cold surroundings whose density (i.e., gas is assumed ideal and the density is for pre-shocked status) profile decreases as a power law $\rho \propto r^{-k}$ where r being varying radius of shock expansion from point blast. They all used the conservation of energy approach to obtain the scaling of the shock radius as a function of time. That is why such solutions are called *first-kind* (or *type*) solutions. Yet in contrast looking at Guderley [26] (also see the discussion in Zel'dovich and Raizer) [25] on implosion problem where they also found a self-similar solution describing imploding shock waves in a constant density environment, energy consideration cannot be used to deduce the scaling of the shock radius as a function of time. Instead, the scaling of the radius as a function of time must be found by forcing that the solution pass through a singular point of equation and that is why such solutions are considered or called self-similar solutions of the *second kind* (or *type*). Therefore, it is safe to say that the second-type solutions do not obey global conservation laws. In reality the true problem, therefore, cannot be completely described by a second-type self-similar solution. Those describe only part of the flow, in some region of interest, whereas other regions deviate from the solutions. So in order to prevent any influences on the self-similar part, a sonic point, where the equations are singular, must separate the nonself-similar parts from it. This requirement replaces the energy conservation as means of reducing the scaling of Lorentz factor with radius, i.e., finding m (See Waxman and Shavarts) [45] for a discussion of the non-relativistic case and (Best and Sari) [46] for the relativistic case. Reference 27 shows that if the density falls fast enough ($k > 3$), energy considerations give the wrong scaling. Same reference also showed that solution should be of the second type for $k > 3.26$. A good discussion around the first- and second-type self-similar solution of implosions and explosions containing ultra-relativistic shocks is given by Re'me Sari [47].

Note

Blandford and McKee [48] are using notation of m and Γ as Lorentz factor of the shocked fluid, and they show for their analysis an adiabatic blast wave where they argue an approximate adiabatic similarity solution as part of suggested blast wave variables which is the appropriate choice of similarity for the well-known Sedov–Taylor similarity for a non-relativistic as follows:

$$\xi = (1 - r/R)\Gamma^2 \geq 0$$

where R is the radius of blast from the center. If the total energy contained in shocked fluid remains constant with t representing the time for shock traveling at some characteristic velocity, then

$$\Gamma^2 \propto t^{-3}$$

If considering the more general case, then we can show the above equation in the following form:

$$\Gamma^2 \propto t^{-m}, \quad m > -1$$

This allows us to treat the case when the energy is supplied continuously at a rate proportional to a power of the time.

Their solution is valid when the density of the external medium into which the shock wave propagates varies with the distance r from the origin as r^{-k},

(continued)

(continued)

for $k < 4$. These are first-type self-similar solutions in which the shock Lorentz factor Γ varies as $\Gamma^2 \propto t^{-m}$, where $m = 3 - k$ to ensure energy conservation.

Best and Sari [46] show new second-type self-similar solutions, valid for $k > 5 - 3\sqrt{3} \approx 4.13$, in their paper. In these type of solution, Γ varies as $\Gamma^2 \propto t^{-m}$, with $m = (3 - 2\sqrt{3})k - 4(5 - 3\sqrt{3})$ so that the shock accelerates and the fraction of the flow energy contained in the vicinity of the shock decrease with time.

In self-similar problem of the second kind, the exponent α cannot be found from dimensional considerations or from the conservation laws without solving the equations. In this case the determination of the similarity exponent requires that the ordinary differential equations for the reduced functions be integrated. Examples of self-similar motions of the second kind are the problems of an imploding shock wave and of an impulsive load, both of which will be discussed in Sect. 1.10 of this chapter.

Solutions of specific problems of the second kind show that in all these cases, the initial conditions of the problem contain only one-dimensional parameter with the unit of mass but lacking parameter A. This condition eliminates the possibility of determining the number α from the dimensions of A, which means it would not be possible to construct the dimensionless combination $\xi = r/At^{\alpha}$. However, the dimensions of this parameter α are not dictated by the initial conditions of the problem, but rather are found from the solution of the equations [25].

1.17 Conclusion

Many engineering problems are too complex to find a mathematically closed form of solution for them. In such cases, a type of analysis, which involves the dimensions of the quantities entering the problem, may be useful. This is as what we have described and shown in different examples called *dimensional analysis*. Uses and applications for dimensional analysis include the following:

- To reduce the number of variables to be studied or plotted
- In planning experiments
- In designing engineering models to be studied and in interpreting model data
- To emphasize the relative importance of parameters entering a problem
- To enable units of measurement to be changed from one system to another

The last of these is common, although relatively trivial, application.

In general, dimensional analysis is any mathematical operation, which involves units or dimensions.

References

1. G.I. Barenblatt, *Dimensional Analysis* (Gordon and Breach Science, New York, 1987)
2. Galileo Galilei, Discorsi e Dimostrazioni Matematiche *intorno à due nuoue scienze* Attenenti alla Mecanica & i Movimenti Locali (1638)
3. I. Stewart, *Does God Play Dice* (Penguin, London, 1989), p. 219
4. Tina Komulainen, Helsinki University of Technology, Laboratory of Process Control and Automation. tiina.komulainen@hut.fi
5. J. Sylvan Katz, The Self-Similar Science System. Research Policy **28**, 501–517 (1999)
6. C. Judd, Fractals ¨C Self-Similarity. http://www.bath.ac.uk/¡«ma0cmj/FractalContents.html. Accessed 16 Mar 2003
7. N. Shiode, M. Batty, Power law distributions in real and virtual worlds. http://www.isoc.org/inet2000/cdproceedings/2a/2a_2.htm. Accessed 17 Mar 2003
8. Charles Stuart University. The Hausdorff Dimension http://life.csu.edu.au/fractop/doc/notes/chapt2a2.html. Accessed 17 Mar 2003
9. P. Krugman, *The Self-Organizing Economy* (Blackwell, Oxford, 1996)
10. G.I. Barenblatt, *Scaling Phenomena in Fluid Mechanics*, 1st edn. (Cambridge University Press, Cambridge, 1994)
11. O. Petruk, *Approximations of the self-similar solution for blast wave in a medium with power-law density variation* (Institute for Applied Problems in Mechanics and Mathematics NAS of Ukraine, Lviv, 2000)
12. G.I. Taylor, Proc. R. Soc. London **A201**, 159 (1950)
13. L. Sedov, Prikl. Mat. Mekh. **10**, 241 (1946)
14. L. Sedov, *Similarity and Dimensional Methods in Mechanics* (Academic, New York, 1959)
15. R.K. Merton, The Matthew effect in science. Science **159**(3810), 56–63 (1968)
16. R.K. Merton, The Matthew effect in science II. ISIS **79**, 606–623 (1988)
17. W.B. Krantz, *Scaling Analysis in Modeling Transport and Reaction Processes* (Wiley Interscience, Hoboken, 1939)
18. I. Proudman, J.R.A. Pearson, J. Fluid Mech **2**, 237 (1957)
19. Harald Hance Olsen. Bukingham's Pi Theorem. www.math.ntnu.no/~hanche/notes/buckingham/buckingham-a4.pdf
20. G.I. Taylor, The formation of a blast wave by a very intense explosion. I. Theoretical discussion. Proc. Roy Soc. A **201**, 159–174 (1950)
21. G.I. Taylor, The formation of a blast wave by a very intense explosion. II. The atomic explosion of 1945. Proc. Roy Soc. A **201**, 175–186 (1950)
22. C.L. Dym, Principle of Mathematical Modeling. 2nd edn, (Elsevier, 2004)
23. L.D. Landau, E.M. Lifshitz, *Fluid Mechanics* (Pergamon Pres, New York, 1959). § 61
24. R. Illner, C.S. Bohun, S. McCollum, T. Van Roode, *Mathematical Modeling, A Case Studies Approach* (American Mathematical Society (AMS), 2005)
25. Y.B. Zel'dovich, Y.P. Raizer, *Physics of Shock Waves and High-Temperature Hydrodynamics Phenomena* (Dover Publication, New York, 2002)
26. G. Guderley, Starke kugelige und zylindrische Verdichtungsstosse in der Nahe des Kugelmittelpunktes bzw. der Zylinderachse. Luftfahrt-Forsch **19**, 302–312 (1942)
27. D.S. Butler, Converging spherical and cylindrical shocks. Armament Research Establishment report 54/54 (1954)
28. J.H. Lau, M.M. Kekez, G.D. Laugheed, P. Savic, Spherically Converging Shock Waves in Dense Plasma Research, in *Proceeding of 10th International Symposium on Shock Tube and Waves*, 1979, p. 386
29. J. Nuckolls, L. Wood, A. Thiessen, G. Zimmerman, Laser compression of matter to super-high densities: thermonuclear (CRT) applications. Nature **239** (1972)
30. I.I. Glass, D. Sagie, Application of Explosive-Driven Implosions to Fusion. Physics of Fluids **25**, 269–270 (1982)

31. F. Winterberg, *The Physical Principles of Thermonuclear Explosive Devices* (Fusion Energy Foundation Frontiers of Science Series, New York, 1981)
32. L. Woljer, Supernova Remnants. Ann. Rev. Astr. **10**, 129 (1972)
33. I.I. Glass, S.P. Sharma, Production of Diamonds From Graphite using Explosive-Driven Implosions. AIAA Journal **14**(3), 402–404 (1976)
34. K.P. Stanyukovich, *Unsteady Motion of Continuous Media* (Academic, New York, 1960)
35. S. Ashraf, Approximate Analytic Solution of Converging Spherical and Cylindrical Shocks with Zero Temperature Gradient in the Rear Flow Field. Z. angew. Math. Phys. **6**(6), 614 (1973)
36. S. Yadegari, Self-similarity. http://www-crca.ucsd.edu/¡«syadegar/MasterThesis/node25.html. Accessed 16 Mar 2003
37. B.J. Carr, A.A. Coley, Self-similarity in general relativity. http://users.math.uni-potsdam.de/~oeitner/QUELLEN/ZUMCHAOS/selfsim1.htm. Accessed 16 Mar 2003
38. V.J. Skoglund, *Similitude, Theory and Applications* (International Textbook Company, Scranton, 1967)
39. H. Schlichting, *Boundary Layer Theory*, 4th edn. (McGraw-Hill Book Company, New York, 1960)
40. G.I. Barenblatt, *'Scaling' Cambridge Texts in Applied Mathematics* (Cambridge University Press, Cambridge, 2006)
41. A. Sakurai, On the problem of shock wave arriving at the edge of gas. Commun. Pure. Appl. Math **13**, 353 (1960)
42. P.L. Sachdev, S. Ashraf, Strong shock with radiation near the surface of a star. Phys. Fluids **14**, 2107 (1971)
43. J. von Neumann, *Blast Waves Los Alamos Science Laboratory Technical Series*, vol. 7 (Los Alamos, 1947).
44. L.I. Sedov, *Similarity and Dimensional Methods in Mechanics* (Academic, New York, 1969). Chap. IV
45. E. Waxman, D. Shvarts, Second-type self-similar solutions to the strong explosion problem. Phys. Fluids A **5**, 1035 (1993)
46. P. Best, R. Sari, Second-type self-similar solutions to the ultra-relativistic strong explosion problem. Phys. Fluids **12**, 3029 (2000)
47. R.'m. Sari, First and second type self-similar solutions of implosions and explosions containing ultra relativistic shocks. Physics of Fluids **18**, 027106 (2006)
48. R.D. Blandford, C.F. McKee, Fluid dynamics of relativistic blast waves. Phys. Fluids **19**, 1130 (1976)

Chapter 2
Dimensional Analysis: Similarity and Self-Similarity

One important aspect of dimensional analysis that is pushing this subject beyond simple Buckingham or Pi-theorem dealing with certain unique problem in gas dynamics, fluid mechanics, or plasma physics and other science- or engineering-related field is implementation similarly and self-similarity. Utilization of similarity in particular is the handling of very complex partial differential equations by converting them to a very simple type of ordinary differential equations, where we can in most cases solve them analytically. Analyses of these equations and seeking an exact solution for them require associated boundary conditions, where these boundary conditions for these partial differential equations are behaving asymptotically, and then finding such exact solution analytically becomes almost very straightforward, and self-similarity method is a good tool to implement.

2.1 Lagrangian and Eulerian Coordinate Systems

Before we go forward with subject of dimensional analysis and utilization of similarity or self-similarity, we have to pay attention to coordinate systems that are, known to engineers and scientists as either Lagrangian or Eulerian coordinate systems. In dealing with the complexity of partial differential equation and quest for their exact solutions analytically, one needs certain defined boundary condition that describes the problem in hand. These boundary conditions need to be defined either in Eulerian or in Lagrangian coordinate system when time is varying for the problem of interest. Therefore, we need to have a grasp of Lagrangian and Eulerian coordinate systems as well as the difference between them.

To have a concept for time t, we need a motion, and motion is always determined with respect to some reference system known as coordinate system in three dimensions. A correspondence between numbers and points in space is established with the aid of a coordinate system. For three-dimensional space, we assume three numbers x_1, x_2, and x_3 correspond to points as three components of X, Y, and Z in

B. Zohuri, *Dimensional Analysis Beyond the Pi Theorem*,
DOI 10.1007/978-3-319-45726-0_2

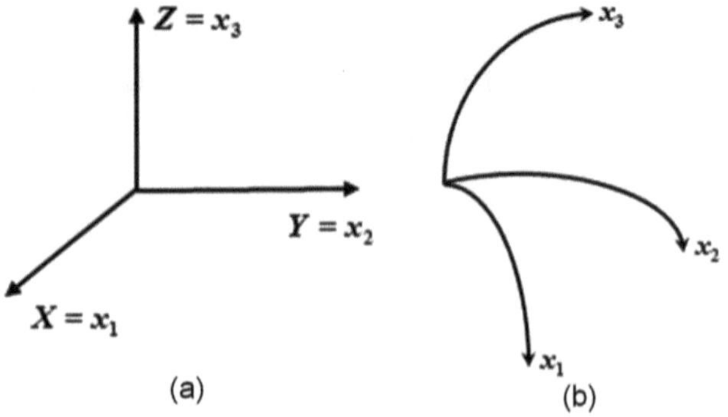

Fig. 2.1 Cartesian (**a**) and curvilinear (**b**) coordinate systems

Cartesian coordinate system and accordingly for Curvilinear coordinate system for its own designated components according to Fig. 2.1a, b, and they are called the coordinates of the point.

In Fig. 2.1a, b, lines along which any two coordinates remain constant are called coordinate lines. For example, the line for which x_2 is constant and x_3 is constant, defines the coordinate line x_1, along which different points are fixed by the values of x_1; the direction of increase of the coordinate x_1 defines the direction along this line. Three coordinate lines may be depicted through each point of space. However, for each point, the tangents to the coordinate lines do not lie in one plane, and, in general, they form a non-orthogonal trihedron. Now assuming these three points mathematically are presented as x_i for $i = 1, 2, 3$ and the coordinate lines x_i are straight, then the system of coordinates is rectilinear and if not, then the system is curvilinear. For our purpose of discussion on the subject of motion of a continuum in this book, it is necessary to present the curvilinear coordinate system, which is essential in continuum mechanics.

Now that we start our introduction with concept of time t, we need to make a notation of time and coordinate system x_i. Therefore, the symbols x_1, x_2, and x_3 will denote coordinates in any system which may also be Cartesian: the symbols of X, Y, and Z in orthogonal form are presented in the Cartesian coordinate system, while the fourth dimension time is designated with the symbol t. Thus, if a point moves relative to the coordinate system x_1, x_2, and x_3, while its coordinate changes in time, then we can mathematically present motion of point as follows:

$$x_i = f_i(t) \quad \text{For} \quad i = 1, 2, 3 \tag{2.1}$$

With this notation, the motion of point will be known if one knows the characteristic and behavior of Eq. 2.1, providing that the moving point coincides with different points of space at different instants of time. This is referred to as the law of motion of the point, and by knowing this law, we can now define motion of a continuum. A continuous medium represents a continuous accumulation of points, and by definition, knowledge of the motion of a continuous medium result in knowledge of the motion of all points. Thus, in general, as one can see, the study of the motion of a volume of a continuous body as a whole is insufficient proposition.

For the above situation, one must treat each distinct point individually in order to form a geometrical point of view that is completely identical points of the continuum. This is referred to as individualization of the points of a continuum, and shown below is how this law is used in theory and is determined by the fact that the motion of each point of a continuous medium is subject to certain physical laws [1].

Let the coordinates of points at the initial time t_0 be denoted by ξ_1, ξ_2, and ξ_3 or for that matter denoted by ξ_i for $i = 1, 2, 3$ and the coordinated of points at an arbitrary instant of time t by x_1, x_2, and x_3 or in general noted as x_i for $i = 1, 2, 3$ as we have done it before. For any point of a continuum, specified by the coordinates ξ_1, ξ_2, and ξ_3, one may write down the law of motion which contains not only functions of a single variable, as in the case of the motion of a point, but of four variables (i.e., all three coordinates plus time): therefore, the initial coordinates ξ_1, ξ_2, and ξ_3 and the time t can be written as

$$\begin{cases} x_1 = x_1(\xi_1, \xi_2, \xi_3, t) \\ x_2 = x_2(\xi_1, \xi_2, \xi_3, t) \\ x_3 = x_3(\xi_1, \xi_2, \xi_3, t) \end{cases} \Rightarrow x_i = x_i(\xi_1, \xi_2, \xi_3, t) \tag{2.2}$$

If in Eq. 2.2 ξ_1, ξ_2, and ξ_3 are fixed and t varies (Eulerian), then Eq. 2.2 describes the law of motion of one selected point of the continuum. If ξ_1, ξ_2, and ξ_3 vary and the time t is fixed, then Eq. 2.2 gives the distribution of the points of the medium in space at a given instant of time (Lagrangian). If ξ_1, ξ_2, and ξ_3 including time t vary, then one may interpret Eq. 2.2 as a formula which determines the motion of the continuous medium, and, by definition, the functions in Eq. 2.2 yield the law of motion of the continuum. The coordinates ξ_1, ξ_2, and ξ_3 or sometimes definite functions of these variables, which individualize the points of a medium, and the time t are referred to as Lagrangian coordinates. In case of continuum mechanics, the fundamental problem is to determine the functions presented in Eq. 2.2.

To expand the above discussions into fluid mechanics in order to analyze fluid flow, different viewpoints can be taken, very similar to using different coordinate systems. For this matter, two different points of view will be discussed for describing fluid flow. They are called Lagrangian and Eulerian viewpoints.

1. *Lagrangian viewpoint*

   The flow description via the Lagrangian viewpoint is a view in which a fluid particle is followed. This point of view is widely used in dynamics and statics and easy to use for a single particle. As the fluid particle travels about the flow field, one needs to locate the particle and observe the change of properties (Fig. 2.2).

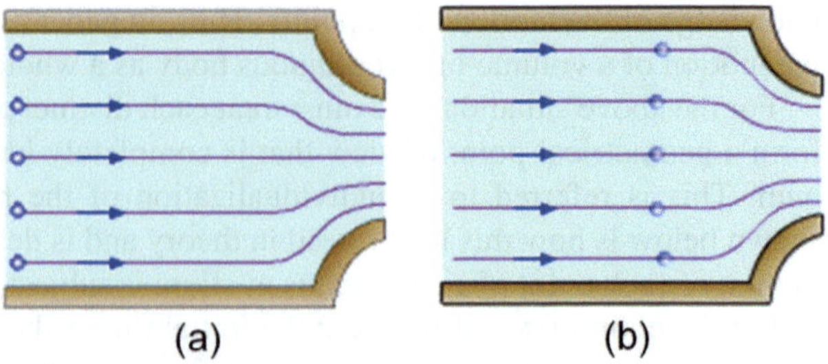

Fig. 2.2 Lagrangian viewpoint

Fig. 2.3 Lagrangian examples. (**a**) Tracking of whales and (**b**) weather balloon

That is,

$$\begin{cases} \vec{r}(t) & \text{Position} \\ T(\xi_1, \xi_2, \xi_3, t) & \text{Temperature} \\ \rho(\xi_1, \xi_2, \xi_3, t) & \text{Density} \\ P(\xi_1, \xi_2, \xi_3, t) & \text{Pressure} \end{cases} \tag{2.3}$$

where ξ_1, ξ_2, and ξ_3 represent a particular particle or object. One example of Lagrangian description is the tracking of whales (position only). In order to better understand the behavior and migration routes of the whales, they are commonly tagged with satellite-linked tags to register their locations, diving depths, and durations (Fig. 2.3).

Another example of Lagrangian system can be thought of as weather balloons free to follow the wind and record data at different locations at that given moment.

2. *Eulerian viewpoint*
The first approach to describe fluid flow is through the Eulerian point of view. The Eulerian viewpoint is implemented by selecting a given location in a flow field (x_1, x_2, x_3) and observing how the properties (e.g., velocity, pressure, and

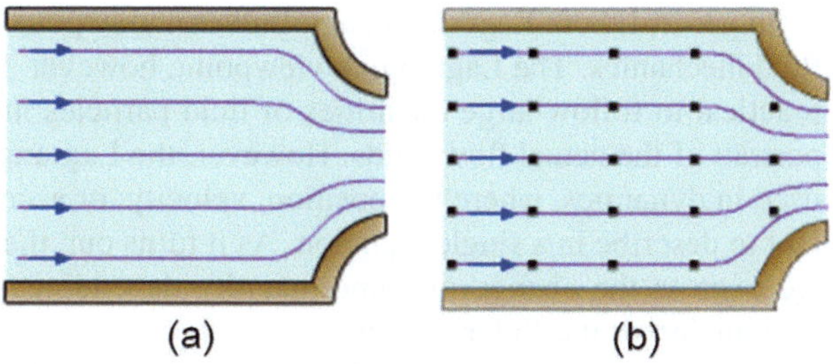

Fig. 2.4 Eulerian viewpoint

Fig. 2.5 Eulerian examples. (**a**) Temperature measurements and (**b**) weather station

temperature) change as the fluid passes through this particular point. As such, the properties at the fixed points generally are functions of time, such as (Fig. 2.4)

$$\begin{cases} \vec{V}(x_1, x_2, x_3, t) & \text{Velocity} \\ T(x_1, x_2, x_3, t) & \text{Temperature} \\ \rho(x_1, x_2, x_3, t) & \text{Density} \\ P(x_1, x_2, x_3, t) & \text{Pressure} \end{cases} \tag{2.4}$$

It should be noted that the position function $\vec{r}(t)$ is not used in Eulerian viewpoint. This is a major difference from the Lagrangian viewpoint, which is used in particle mechanics (i.e., dynamics and statics).

However, if the flow is *steady*, then the properties are no longer function of time. The Eulerian viewpoint is commonly used, and it is the preferred method in the study of fluid mechanics. Take the experimental setup as shown in the figure, for example. Thermocouples (temperature sensors) are usually attached at fixed locations to measure the temperature as the fluid flows over the nonmoving sensor location (Fig. 2.5).

Another intuitive explanation can be given in terms of weather stations. The Eulerian system can be thought as land-based weather stations that record temperature, humidity, etc., at fixed locations at different times.

In general, both Lagrangian and Eulerian viewpoints can be used in the study of fluid mechanics. The Lagrangian viewpoint, however, is seldom used since it is not practical to follow large quantities of fluid particles in order to obtain an accurate portrait of the actual flow fields. However, the Lagrangian viewpoint is commonly used in dynamics, where the position, velocity, or acceleration over time is important to describe in a single equation. As it turns out, there is a big difference in how we express the change of some quantity depending on whether we think in the Lagrangian or the Eulerian sense.

In summary, there are two different mathematical representations of fluid flow:

1. The Lagrangian picture in which we keep track of the locations of individual fluid particles. Picture a fluid flow where each fluid particle carries its own properties such as density, momentum, etc. As the particle advances, its properties may change in time. The procedure of describing the entire flow by recording the detailed histories of each fluid particle is the Lagrangian description. In other words, pieces of the fluid are "tagged." The fluid flow properties are determined by tracking the motion and properties of the particles as they move in time. A neutrally buoyant probe is an example of a Lagrangian measuring device.

   The particle properties at position $\vec{r}(t)$ such as temperature, density, pressure $\cdots$ can be mathematically represented as follows: $T(\xi_i, t)$, $\rho(\xi_i, t)$, $P(\xi_i, t)$, $\cdots$ for $i = 1, 2, 3$. Note that ξ_i is a representation of a fixed point in three-dimensional space at given time t, which may include initial time t_0. The Lagrangian description is simple to understand: conservation of mass and Newton's laws applies directly to each fluid particle. However, it is computationally expensive to keep track of the trajectories of all the fluid particles in a flow, and therefore the Lagrangian description is used only in some numerical simulations.
2. The Eulerian picture in which coordinates is fixed in space (the laboratory frame). The fluid properties such as velocity, temperature, density, pressure, $\cdots$ are written as functions of space and time. The flow is determined by analyzing the behavior of the functions. In other words, rather than following each fluid particle, we can record the evolution of the flow properties at every point in space as time varies. This is the Eulerian description. It is a field description. A probe fixed in space is an example of Eulerian-measuring device.

   This means that the flow properties at a specified location depend on the location and on time. For example, the velocity, temperature, density, pressure, $\cdots$ can be mathematically represented as follows: $\vec{V}(x_i, t)$, $T(x_i, t)$, $\rho(x_i, t)$, $P(x_i, t)$, $\cdots$ for $i = 1, 2, 3$. Note that x_i is the location of fluid at time t.

   The Eulerian description is harder to understand: how do we apply the conservation laws? However, it turns out that it is mathematically simpler to apply. For this reason, in fluid mechanics, we use mainly the Eulerian description.

The aforementioned locations are described in coordinate systems.

2.1.1 Arbitrary Lagrangian–Eulerian (ALE) Systems

The arbitrary Lagrangian–Eulerian that is noted as ALE is a formulation in which computational system is not a priori fixed in space (e.g., Eulerian-based formulation) or attached to material or fluid stream (e.g., Lagrangian-based formulations). ALE-based formulation can alleviate many of the drawbacks that the traditional Lagrangian-based and Eulerian-based formulation or simulation have.

When using the ALE technique in engineering simulations, the computational mesh inside the domains can move arbitrarily to optimize the shapes of elements, while the mesh on the boundaries and interfaces of the domains can move along with materials to precisely track the boundaries and interfaces of a multi-material system.

ALE-based finite element formulations can reduce to either Lagrangian-based finite element formulations by equating mesh motion to material motion or Eulerian-based finite element formulations by fixing mesh in space. Therefore, one finite element code can be used to perform comprehensive engineering simulations, including heat transfer, fluid flow, fluid–structure interactions, and metal manufacturing.

Some applications of arbitrary Lagrangian–Eulerian (ALE) in finite element techniques that can be applied to many engineering problems are:

- Manufacturing (e.g., metal forming/cutting, casting)
- Fluid–structure interaction (combination of pure Eulerian mesh, pure Lagrangian mesh and ALE mesh in different regions)
- Coupling of multi-physics fields with multi-materials (moving boundaries and interfaces)

Another important application of ALE is the particle-in-cell analyses, particularly plasma physics. The particle-in-cell (PIC) method refers to a technique used to solve a certain class of partial differential equations. In this method, individual particles (or fluid elements) in a Lagrangian frame are tracked in continuous phase space, whereas moments of the distribution such as densities and currents are computed simultaneously on Eulerian (stationary) mesh points.

PIC methods were already in use as early as 1955 [2], even before the first FORTRAN compilers were available. The method gained popularity for plasma simulation in the late 1950s and early 1960s by Buneman, Dawson, Hockney, Birdsall, Morse, and others. In plasma physics applications, the method amounts to following the trajectories of charged particles in self-consistent electromagnetic (or electrostatic) fields computed on a fixed mesh [3].

2.2 Similar and Self-Similar Definitions

We need to have a better understanding of similar and self-similar methods and their definition in the subject of dimensional analysis. Once we have these methods defined properly, then we can extend it to motion of a medium in particular from self-similarity point of view. In addition, we are able to deal with complexity of partial differential equations of conservation laws, such as law of conservation of mass, momentum, and, finally, energy, of nonlinear type both in Eulerian and in Lagrangian schemes using all three coordinate systems that we are familiar with. These coordinates are Cartesian, cylindrical, and spherical coordinate systems. Further, this allows us to have a better understanding of what is the self-similarity of first and second kind, and their definitions, what are the differences between them, as well as where and how they get applied to our physics and mathematics problems in hand. Few of these examples that we can mention here are, self-similar motion of a gas with central symmetry, both sudden explosion (Taylor) [4, 5] and sudden implosion (Guderley) [6] problems. The first one is considered self-similarity of first kind, while the latter is considered as self-similarity of second kind. Through these understandings, we can have a better grasp of gas dynamics differential equations and their properties in a medium. In addition, the analysis of such differential equations for a gas motion with central symmetry becomes much easier, by utilizing self-similar method.

Self-similar motion of a medium is one in which the parameters that are characterizing the state and motion of the medium vary in a way as the time varies; the spatial distribution of any of these parameters remains similar to itself. However, the scale characterizing this perturbation/distribution can also vary with time in accordance with definite rules. In other words, if the variation of any of the above parameters with time are specified at a given point in space, then the variation of these parameters with time will remain, the same at other points lying on a definite line or surface, providing that the scale of a given parameter and the value of the time are suitably changed [8].

The analytical conditions for self-similar motion lead to one or more relations between the independent variables, defining functions which play the role of new independent variables using dimensional analysis and self-similarity approach [7]. This approach follows that, in case of self-similar motion, the number of independent variables in the fundamental systems of equations is correspondingly reduced. This technique, considerably, simplifies the complex and nonlinear partial differential equations to sets of ordinary differential equations. Thus, sometimes, this makes it possible to obtain several analytical solutions describing, for example, the self-similar motion of the medium. As it was said, in the case of two independent variables, and sometimes even in the case of three independent variables, the fundamental system of equations becomes a system of ordinary rather than partial differential equations [8].

Applications of self-similar approach can be seen to all unsteady self-similar motions with symmetry, all steady plane motions, and certain axial symmetrical motions as well. These types of approaches have solved problems of self-similarity of first kind [4, 5] and second kind [6] in the past, where complex partial differential equations of conservation laws are described by systems of ordinary differential equations. Investigation of the most important modern gas dynamic motions or plasma physics such as laser-driven pellet for fusion confinement via self-similar methods enables us to produce very useful conclusions by solving the conservation law equations in them, using self-similarity model. To be concerned about the more general types of motion of the medium also allows us to develop and establish laws of motion in various cases of practical interest. They may include the propagation of strong shock waves in case of explosion and implosion events, and propagation of soliton waves and the reflection of shock waves are few examples that can fall into category of self-similarity methods. To further have better understanding of subject similarity and self-similarity requires knowledge of fundamental equation of gas dynamics, where we can investigate a compressible liquid or gas. Therefore, the next few sections of this chapter are allocated to this matter and related thermodynamics aspect of state of medium equations. For this, we also need to understand the difference between compressible and incompressible flows. In addition, the detailed analyses of similarity can be found in the book by this author, so we do not have to repeat the same information here [7].

2.3 Compressible and Incompressible Flows

Compressible flows can be observed in many applications in Aerospace and various topics such as mechanical engineering, fluid mechanics, gas dynamics, etc. Some intuitive examples are flows in nozzles of jet or rocket engines, compressors, turbine and diffuser, as well as study of strong or weak shock waves generated by point blast and other related phenomena. In almost all of these examples, air or some other gas fluid or mixture of gases are the working fluids or medium. Depends on the application in which compressible flow occurs is quite large and hence, to have a better understanding of the dynamics of compressible flow and compressibility phenomenon for engineers and scientist, studying this field is very essential.

All fluids to some extent are compressible in one way or another, and for that, we can define the compressibility of fluid factor as τ, mathematically defined as follows:

$$\tau = -\frac{1}{\upsilon}\frac{\partial \upsilon}{\partial P} \tag{2.5}$$

where

$\upsilon =$ is the specific volume, and

$P =$ is the pressure

One occurrence that is inevitable according Eq. 2.5 is that any change in specific volume results in a given change in pressure accordingly and it will take place per compression process. This also would indicate that for a given change in pressure, the change in specific will be different between an isothermal and adiabatic compression process.

Using our fundamental knowledge of thermodynamics, we can easily define the compressibility of a fluid by expressing the specific volume as a function of temperature T and pressure P as $\upsilon = \upsilon(T, P)$; thus, we can write the following notation utilizing differential calculus of chain rule:

$$d\upsilon = \left(\frac{\partial \upsilon}{\partial P}\right)_T dP + \left(\frac{\partial \upsilon}{\partial T}\right)_P dT \tag{2.6}$$

The subscript T and P in Eq. 2.6 indicates that the temperature and pressure is held constant during the expansion, respectively. Analysis of Eq. 2.6 reveals that the first term on the right-hand side (RHS) of this equation, namely, can define a new term known as *isothermal compressibility* in the form of $-\frac{1}{\upsilon}\left(\frac{\partial \upsilon}{\partial P}\right)_T$ and the second term on RHS of Eq. 6.2 is presentation of *volumetric thermal expansion coefficient*, which is written in form of $\frac{1}{\upsilon}\left(\frac{\partial \upsilon}{\partial P}\right)_P$. This second term basically represents the change in specific volume or equivalent density due to a change in temperature. For instance, when a gas is heated up at constant pressure, the density decreases, and the specific volume increases correspondingly. This change can be large as it can be seen in the case of most combustion systems or equipments, without necessarily having any implications on the compressibility of the fluid. Therefore, a need for following that compressibility effect is required, and it is very important only when the change in specific volume or equivalent density is due largely to a change in pressure.

Equation 2.5 can have a new form in terms of density ρ, as

$$\tau = \frac{1}{\rho}\frac{\partial \rho}{\partial P} \tag{2.7}$$

However, the isothermal compressibility of water and air under standard atmospheric conditions is 5×10^{-10} m^2/N and 10^{-5} m^2/N. Thus, water (in liquid phase) can be treated as an incompressible fluid in all applications. On the contrary, it would seem that air, with a compressibility that is five orders of magnitude higher, has to be treated as a compressible fluid in all applications. Fortunately, this is not true when flow is involved [9].

One way to identify a compressibility issue is the study of sound pressure wave propagation in any medium with a speed, which depends on the bulk compressibility. We can state that the less compressible the medium, the higher the speed of sound. Thus, speed of sound is a convenient source of reference speed, when flow is involved. Speed of sound in air under normal atmospheric conditions is measured to be 330 m/s. However, using speed of sound as reference point requires understanding of another factor known as Mach number, which is a dimensionless quantity and is represented by letter M and states as the ratio of flow velocity u past a boundary to the local speed of sound a and is mathematically written as

$$M = \frac{V}{a} \tag{2.8}$$

where

$M =$ is the Mach number.
$V =$ is the local flow velocity with respect to the boundaries (either internal, such as an object immersed in the flow, or external, like a channel).
$a =$ is the speed of sound in the medium.

In the simplest explanation, the speed of Mach 1 is equal to the speed of sound. Therefore, Mach 0.65 is about 65 % of the speed of sound (subsonic), and Mach 1.35 is about 35 % faster than the speed of sound (supersonic). See Fig. 2.6

The local speed of sound, and thereby the Mach number, depends on the condition of the surrounding medium, in particular, the temperature and pressure. The Mach number is primarily used to determine the approximation with which a flow can be treated as an incompressible flow. The medium can be a gas or a liquid. The boundary can be traveling in the medium, or it can be stationary while the medium flows along it, or they can both be moving, with different velocities: what matters is their relative velocity with respect to each other.

The boundary can be the boundary of an object immersed in the medium or of a channel such as a nozzle, diffusers, or wind tunnels channeling the medium. As the Mach number is defined as the ratio of two speeds, it is a dimensionless number. If $M < 0.2 - 0.3$ and the flow is quasi-steady and isothermal, compressibility effects will be small, and a simplified incompressible flow equations can be used.

Fig. 2.6 An F/A-18 Hornet creating a vapor cone at transonic speed just before reaching the speed of sound

The Mach number is a measurement of a quantity at which an aircraft can fly and easily can be calculated in terms of adiabatic index of refraction γ and that is the ratio of specific heat of a gas at a constant pressure C_P to specific heat of it at a constant volume C_V, namely, $\gamma = C_P/C_V$. By pressure here we mean static pressure P in contrast to impact pressure Q ; this pressure is sometimes called dynamic pressure as well. Therefore, to do this analysis, we start our approach by stating that the dynamic pressure is shown as

$$Q = \frac{\gamma}{2} PM^2 \tag{2.9}$$

Assuming air to be an ideal gas in our case here, the formula to compute Mach number in a subsonic compressible flow is derived from Bernoulli's equation for

$$M = \sqrt{\frac{2}{\gamma - 1}\left[\left(\frac{Q_a}{P} + 1\right)^{\frac{\gamma-1}{\gamma}} - 1\right]} \tag{2.10}$$

where

$Q_a =$ is impact pressure or dynamic pressure due to air

$P =$ is the static pressure

$\gamma =$ is the adiabatic index for the air

The formula to compute Mach number in a supersonic compressible flow (i.e., dynamic pressure P at supersonic condition) is derived from the Rayleigh supersonic formula:

$$\frac{P_t}{P} = \left[\frac{\gamma - 1}{2} M^2\right]^{\left(\frac{\gamma-1}{\gamma}\right)} \cdot \left[\frac{\gamma + 1}{\left(1 - \gamma + 2\gamma M^2\right)}\right]^{\left(\frac{1}{\gamma-1}\right)} \tag{2.11}$$

In summary, we can up with a quantitative criterion to give us an idea about the importance of compressibility effects in the flow by using simple scaling arguments as follows. From Bernoulli's equation for steady flow, it follows that $\Delta P\tilde{\ }\rho U^2$, where U is some characteristic speed. It can be demonstrated that speed of sound $a = \sqrt{\Delta P/\Delta \rho}$, wherein ΔP and $\Delta \rho$ correspond to an isentropic process. Thus,

$$\frac{\Delta \rho}{\rho} = \frac{1}{\rho}\frac{\Delta \rho}{\Delta P}\Delta P = \frac{1}{\rho}\left(\frac{1}{a^2}\right)\left(\rho U^2\right) = \frac{U^2}{a^2} = M^2 \tag{2.12}$$

On the other hand, upon rewriting Eq. 2.7 for an isentropic process, we get the following:

$$\frac{\Delta\rho}{\rho} = \tau_{\text{isentropic}}\Delta P \tag{2.13}$$

Comparison of these two equations, namely, Eqs. 2.12 and 2.13, reviles clearly that, in the presence of a flow, density changes are proportional to the square of the Mach number.

Note that this is true for steady flows only. For unsteady flows, density changes are proportional to the Mach number.

It is customary to assume that the flow is essentially incompressible if the change in density is less than 10 % of the mean value, providing the change is predominantly due to a change in pressure. It thus follows that compressibility effects are significant only when the Mach number exceeds 0.3. To prove this point, we go through following the analyses.

2.3.1 Limiting Condition for Compressibility

From definition of incompressible flow, we can state that, as long as gas flows at a sufficiently low speed from one cross section to another, the change in volume or density can be neglected, and, therefore, the flow can be treated as in compressible flow. Although the fluid is compressible, this property may be neglected when the flow is taking place at low speed. In other words, although there is some density change associated with every physical flow, it is often possible for low speed flows to neglect it and to idealize the flow as incompressible. A practical application of this approximation at low speed is for flow around an airplane or, for another virtual example we can look at, flow through a vacuum cleaner.

From the above discussion, it is clear that compressibility is the phenomenon by virtue of which the flow changes its density with change in speed. Now to be able to address the question under what precise conditions density changes must be taken under consideration. The answer can be found in a quantitative measure of compressibility, where volume modules of elasticity $\mathbb{E}$ can be defined as follows [10]:

$$\Delta P = -\mathbb{E}\frac{\Delta\mathbb{V}}{\mathbb{V}_{\text{i}}} \tag{2.14}$$

where

ΔP = is change in static pressure
$\Delta\mathbb{V}$ = is change in volume
$\mathbb{V}_{\text{i}}$ = is the initial volume

Now, for an ideal gas, the pressure can be expressed by the equation of state as

$$P = \rho RT \tag{2.15}$$

where

$P =$ is static pressure
$T =$ is the temperature
$R =$ is the ideal gas constant
$\rho =$ is the gas density

Under isothermal process, in particular, we can write

$$P\mathbb{V} = P_{\mathrm{i}}\mathbb{V}_{\mathrm{i}} = \text{constant} \tag{2.16}$$

where P_{i} is the initial static pressure. Equation 2.16 for small variation in pressure and volume can be expanded into the following form as

$$(P_{\mathrm{i}} + \Delta P)(\mathbb{V}_{\mathrm{i}} + \Delta\mathbb{V}) = P_{\mathrm{i}}\mathbb{V}_{\mathrm{i}} \tag{2.17}$$

Expanding Eq. 2.17 and neglecting the second-order term of (), we get the following result as

$$\begin{aligned} &P_{\mathrm{i}}\mathbb{V}_{\mathrm{i}} + P_{\mathrm{i}}\Delta\mathbb{V} + \Delta P\mathbb{V}_{\mathrm{i}} + \cancel{\Delta P\Delta\mathbb{V}} = P_{\mathrm{i}}\mathbb{V}_{\mathrm{i}} \\ &\Delta P\mathbb{V}_{\mathrm{i}} + \Delta\mathbb{V}P_{\mathrm{i}} = 0 \end{aligned} \tag{2.18}$$

Therefore,

$$\Delta P = -P_{\mathrm{i}}\frac{\Delta\mathbb{V}}{\mathbb{V}_{\mathrm{i}}} \tag{2.19}$$

Comparing Eq. 2.14 with Eq. 2.19, for the gases, we can write

$$\mathbb{E} = P_{\mathrm{i}} \tag{2.20}$$

Therefore, by virtue of Eq. 2.19, the compressibility may be defined as the volume modulus of the pressure.

Limiting conditions for compressibility could be expressed by utilizing conservation of mass concept, where we have

$$\dot{m} = \rho V = \text{constant} \tag{2.21}$$

where

$\dot{m}\cdot =$ is mass flow rate per unit area
$V =$ is the flow velocity
$\rho =$ is the corresponding density of the fluid

Equation 2.21 can also be expanded in a different form for small change in velocity and density at some initial point i as below:

$$(V_{\mathrm{i}} + \Delta V)(\rho_{\mathrm{i}} + \Delta\rho) = \rho_{\mathrm{i}} V_{\mathrm{i}} \tag{2.22}$$

Expanding Eq. 2.22 and neglecting the second-order term of $(\Delta\rho\Delta V)$ simplifies to the following form:

$$\begin{aligned} &\rho_{\mathrm{i}} V_{\mathrm{i}} + \Delta\rho V_{\mathrm{i}} + \rho_{\mathrm{i}} \Delta V + \cancel{\Delta\rho\Delta V} = \rho_{\mathrm{i}} V_{\mathrm{i}} \\ &\Delta\rho V_{\mathrm{i}} + \rho_{\mathrm{i}} \Delta V = 0 \\ &\frac{\Delta\rho}{\rho_{\mathrm{i}}} = -\frac{\Delta V}{V_{\mathrm{i}}} \end{aligned} \tag{2.23}$$

Substituting the result of Eq. 2.23 into Eq. 2.14 and noting that $V = \mathbb{V}$ (i.e., velocity = volume) for unit area per unit time in the present case, we get

$$\Delta P = \mathbb{E} \frac{\Delta\rho}{\rho_{\mathrm{i}}} \tag{2.24}$$

From Eq. 2.24, it is seen that the compressibility may also be defined as the density modulus of the pressure.

For incompressible flows, from Bernoulli's equation,

$$\boxed{P + \frac{1}{2}\rho V^2 = \text{Constant} = P_{\text{stagnation}}} \tag{2.25}$$

The above equation may also be written as

$$P_{\text{stagnation}} - P = \Delta P = \frac{1}{2}\rho V^2 \tag{2.26}$$

Hence, the change in pressure is $\frac{1}{2}\rho V^2$. Using Eq. 2.24 in the above relation, we obtain the following result as

$$\frac{\Delta P}{\mathbb{E}} = \frac{\Delta\rho}{\rho_{\mathrm{i}}} = \frac{\rho_{\mathrm{i}} V_{\mathrm{i}}^2}{2\mathbb{E}} = \frac{Q_{\mathrm{i}}}{\mathbb{E}} \tag{2.27}$$

where $Q_{\mathrm{i}} = \frac{1}{2}\rho_{\mathrm{i}} V_{\mathrm{i}}^2$ is the dynamic pressure as before. Equation 2.2 relates the density change with flow speed.

The compressibility effects can be neglected if the density changes are very small, i.e., if

$$\frac{\Delta\rho}{\rho_i} \ll 1 \tag{2.28}$$

From Eq. 2.27, it is seen that for neglecting compressibility,

$$\frac{Q_i}{\mathbb{E}} \ll 1 \tag{2.29}$$

For gases, the speed of sound wave propagation was noted as symbol of a may be expressed in terms of pressure and density as

$$a^2 = \frac{\Delta P}{\Delta\rho} \tag{2.30}$$

Equation 2.30 is also known as Laplace's equation and is valid for any fluid.

Using Eq. 2.24 in the above relation, we get

$$a^2 = \frac{\mathbb{E}}{\rho_i} \tag{2.31}$$

With this, Eq. 2.27 reduces to the following form as

$$\frac{\Delta\rho}{\rho_i} = \frac{\rho_i}{2}\frac{V_i^2}{\mathbb{E}} = \frac{1}{2}\left(\frac{V}{a}\right)^2 \tag{2.32}$$

The ratio V/a as we know from before called Mach number M. Therefore, the condition of incompressibility for gases based on compression between Eq. 2.32 and Eq. 2.28 is established as

$$\frac{M^2}{2} \ll 1 \tag{2.33}$$

Thus, the criterion which determines the effect of compressibility for gases is M. It is widely accepted that compressibility can be neglected when

$$\frac{\Delta\rho}{\rho_i} \leq 0.05 \tag{2.34}$$

Therefore, $M \leq 0.3$. In other words, the flow is treated as incompressible when $V \leq 100\ \text{m/s}$, i.e., when $V \leq 360\ \text{km/h}$ under standard conditions of the air. The above values of Mach number M and velocity V are widely accepted values, and they may be refixed at different levels, depending upon the flow situation and the degree of accuracy desired.

2.4 Mathematical and Thermodynamic Aspect of Gas Dynamics

Thermodynamic concepts and relations such as first law of thermodynamics (energy equation) and the second law of thermodynamics (entropy equation) play a substantial part in the study of gas dynamics. The system of equations explaining the motion and describing properties of the medium includes the equation of state of the medium, which is one of the fundamental thermodynamic equations. Thus, it is appropriate to express some of the mathematical aspect of these fundamental equations.

2.4.1 *First Law of Thermodynamics*

It is the generalization of the principle of conservation of energy to include energy transfer through heat as well as mechanical work. This law can associate these quantities in the following form:

$$U_2 - U_1 = \Delta U = Q - W \tag{2.35}$$

In this equation, the symbol of U represents internal energy; and during a change of state of the system, the internal energy may change from an initial value U_1 to a final value U_2 and note that such change in internal energy is shown as $U_2 - U_1 = \Delta U$. Heat transfer is energy transfer that is adding a quantity of heat Q to a system, without doing any work during the process, which will increase the internal energy by an amount equal to $Q = \Delta U$. When a system does work which is shown by the symbol W, by expanding against its surrounding and no heat is added during the process, energy leaves the system and the internal energy decreases. That is, when $W > 0$, then $\Delta U < 0$ and vice versa. When both heat transfer and work occur, the total change in internal energy is stated by Eq. 2.35 and can be rearranged to the following form.

$$Q = \Delta U + W \tag{2.36}$$

In an isolated system, when one does not do any work on its surroundings and has no heat flow to or from its surroundings, then $W = Q = 0$, which results in $U_2 - U_1 = \Delta U = 0$, and this is an indication of the fact that *the internal energy of an isolated system is constant.*

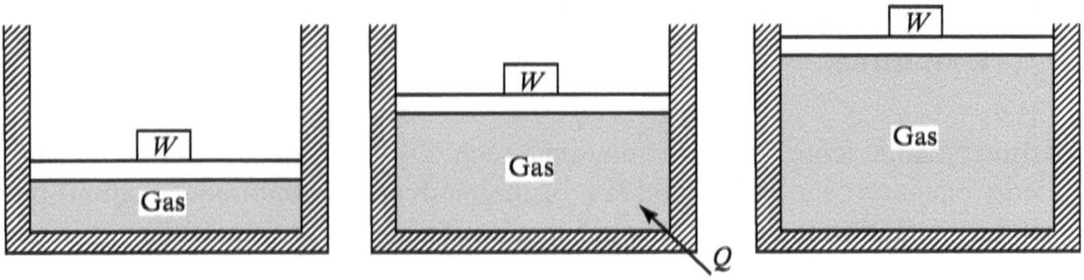

Fig. 2.7 Constant pressure heat conduction

2.4.2 The Concept of Enthalpy

In the solution of problems involving systems, certain products or sums of properties occur with regularity. One such combination of properties can be, demonstrated by considering the addition of heat to the constant pressure situation as it is, shown in Fig. 2.7.

Heat is added slowly to the system (the gas in the cylinder), which is maintained at constant pressure by assuming a frictionless seal between the piston and the cylinder. If the kinetic energy changes and potential energy changes of the system are neglected and all other work modes are absent, the first law of thermodynamics requires that Eq. 2.35 applies

$$W - Q = U_2 - U_1 \tag{2.37}$$

The work done using the weight for the constant pressure P and initial volume V_1 to final volume V_2 process is given by

$$W = P(V_2 - V_1) \tag{2.38}$$

Then, the first law of thermodynamics results in

$$Q = (U + PV)_2 - (U + PV)_1 \tag{2.39}$$

The quantity in parentheses $U + PV$ is a combination of properties, and it is thus a property itself. It is called the *enthalpy* of the system and presented by symbol of H as

$$H = U + PV \tag{2.40}$$

The specific enthalpy h is found by dividing Eq. 2.40 by the unit mass and is written as

$$h = u + Pv \tag{2.41}$$

Enthalpy is a property of a system. It is so useful that it is tabulated in the steam tables along with specific volume and specific internal energy in a book by Zohuri and McDaniel [11]. The energy equation can now be written for a constant pressure process as

$$Q_{1-2} = H_2 - H_1 \tag{2.42}$$

The enthalpy has been defined assuming a *constant pressure system* with difference in enthalpies between two states being in heat transfer. For a variable pressure process, the difference in enthalpy is not quite as obvious. However, enthalpy is still of use in many engineering problems, and it remains a property as defined by Eq. 2.40. In non equilibrium, *constant pressure process* ΔH would not equal the heat transfer.

Because only *changes* in the enthalpy or the internal energy are important, the datum for each can be chosen arbitrarily. Normally the saturated liquid at 0 °C is chosen as the datum point for water [11].

2.4.3 Specific Heats

For a simple system, only two independent variables are necessary to establish the state of the system as specified by Gibbs phase rule. This means that properties like specific internal energy u can be tabulated as a function of two variables In the case of u, it is particularly useful to choose T and v. Or

$$u = u(T, v) \tag{2.43}$$

Using the calculus of chain rule, we express the differential in terms of the partial derivative as follows using the function u and its independent variables T and v in Eq. 2.43;

$$du = \left(\frac{\partial u}{\partial T}\right)_v dT + \left(\frac{\partial u}{\partial v}\right)_T dv \tag{2.44}$$

Since u, v, and T are all properties, the partial derivatives is also a property and is called the *constant volume-specific heat*, shown as C_v and that is

$$C_v \left(\frac{\partial u}{\partial T}\right)_v \tag{2.45}$$

Since many experiments have shown that when a low-density gas undergoes a free expansion its temperature does not change such a gas is by definition an ideal gas. The conclusion is that the internal energy of an ideal gas depends only on its temperature, not on its pressure or volume. This property, in addition to the ideal gas equation of state, is part of the ideal gas model. Because there is no change in temperature, there is no net heat transfer to the substance under experiment.

Obviously since no work is involved, the first law requires that the internal energy of an ideal gas does not depend on volume. Thus, the second term in Eq. 2.44 is zero [11]:

$$\left(\frac{\partial u}{\partial \upsilon}\right)_T = 0 \tag{2.46}$$

Combining Eqs. 2.44, 2.45, and 2.46 will conclude the following final for m for internal energy element *du* as

$$du = C_\upsilon dT \tag{2.47}$$

Equation 2.47 can be integrated as follows:

$$u_2 - u_1 = \int_{T_1}^{T_2} C_\upsilon dT \tag{2.48}$$

For a known C_υ, Eq. 2.48 can be integrated to find the change in internal energy over any temperature interval for an ideal gas. By a similar argument, considering specific enthalpy to be dependent on the two variables, temperature T and pressure p, we have

$$h = h(T, P) \tag{2.49}$$

Note for conveniece of similarity between two specific heats at constant volume and pressure, we are writing lower case for pressure p than uppercase.

Similarly, Eq. 2.49 with usage of calculus of chain rule, provides

$$dh = \left(\frac{\partial h}{\partial T}\right)_p dT + \left(\frac{\partial h}{\partial p}\right)_T dp \tag{2.50}$$

The *constant pressure-specific heat* C_P is defined as the partial

$$C_P = \left(\frac{\partial h}{\partial T}\right)_p \tag{2.51}$$

For an ideal gas and the definition of enthalpy as per Eq. 2.41 in combination with Eq. 2.15 for unit mass, where specific volume is equal to $\upsilon = (1/\rho)$, we can write

$$h = u + p\upsilon = u + RT \tag{2.52}$$

In establishing Eq. 2.52, we have used the ideal gas equation of state. Since T is only a function of T per Eq. 2.49, h is also only a function of T for an ideal gas. Hence, for an ideal gas

$$\left(\frac{\partial h}{\partial p}\right)_T = 0 \tag{2.53}$$

Then from Eq. 2.50, we have

$$dh = C_p dT \tag{2.54}$$

Integrating Eq. 2.54 over the temperature range T_1 to T_2 will give the following result

$$h_2 - h_1 = \int_{T_1}^{T_2} C_p dT \tag{2.55}$$

Often it is convenient to denote the specific heat on a per-mole rather than a per-kilogram basis and the specific heats [11].

The enthalpy Eq. 2.41 or Eq. 2.52 for an ideal gas can be written as

$$dh = du + d(pv) \tag{2.56}$$

Introducing the specific heat relations of Eq. 2.47 and Eq. 2.54 as well as the ideal gas equation gives

$$C_P dT = C_v dT + RdT \tag{2.57}$$

Dividing both sides by dT results in the following relation for an ideal gas:

$$\begin{aligned} C_p &= C_v + R \\ R &= C_p - C_v \end{aligned} \tag{2.58}$$

Note that the difference between C_p and C_v for an ideal gas is always a constant, even though both are functions of temperature. Defining the ratio of specific heats γ, we can see this ratio is also a property of interest and is written as

$$\gamma = \frac{C_p}{C_v} \tag{2.59}$$

Substituting Eq. 2.59 into Eq. 2.58 results in the following useful relationships:

$$C_P = \left(\frac{\gamma}{\gamma - 1}\right) R \tag{2.60}$$

or

$$C_v = \left(\frac{1}{\gamma - 1}\right)R \tag{2.61}$$

Since R for an ideal gas is constant, the specific heat ratio γ just depends on temperature T.

For gases, the specific heat slowly increases with increasing temperature. Since they do not vary significantly over fairly large temperature differences, it is often acceptable to treat C_p and C_v as constants. In this case, the integration is simple, and the internal energy and enthalpy can be expressed as [11]

$$u_2 - u_1 = C_v(T_2 - T_1) \tag{2.62}$$

and

$$u_2 - u_1 = C_v(T_2 - T_1) \tag{2.63}$$

2.4.4 Speed of Sound

Sound waves are infinitely small pressure disturbances. The speed with which sound propagates in a medium is called *speed of sound* and is denoted by the symbol a. Here we are not going to show the derivation of speed of sound, except writing few relationships of it. As it was stated previously, the Laplace's equation of sound that is valid for any fluid is expressed as

$$a^2 = \frac{dp}{d\rho} \tag{2.64}$$

The sound wave is a weak compression wave, across which only infinitesimal change in fluid properties occurs. Furthermore, the wave itself is extremely thin, and changes in properties occur very rapidly. The rapidity of the process rules out the possibility of any heat transfer between the system of fluid particles and its surroundings.

For very strong pressure waves, the traveling speed of disturbance may be greater than that of sound. The pressure can be expressed as

$$p = p(\rho) \tag{2.65}$$

For isentropic process of a gas,

$$\boxed{\frac{p}{\rho^{\gamma}} = \text{constant}} \tag{2.66}$$

Where the isentropic index γ is the ratio of specific heats and is a constant for a perfect gas. Using the above relation in Eq. 2.64, we get

$$a^2 = \frac{\gamma p}{\rho} \tag{2.67}$$

For a perfect gas, by the state equation as we stated in Eq. 2.15,

$$p = \rho RT \tag{2.68}$$

where, again, R is the gas constant and T the static temperature of the gas in absolute units.

Equations 2.67 and 2.68 together lead to the following expression for the speed of sound:

$$a = \sqrt{\gamma RT} \tag{2.69}$$

The assumption of perfect gas is valid so long as the speed of gas stream is not too high. However, at hypersonic speeds, the assumption of perfect gas is not valid, and we must consider Eq. 2.67 to calculate the speed of sound.

Implication of variation of speed of sound a with altitude is explained in the following example.

Example 2.1 For an aircraft flying at speed of 1000 km/h, the variation of speed of sound a, and Mach number M with altitude is as follows:

From the International Standard Atmosphere (ISA), $T = 15°\text{C}$ at sea level. Therefore, the speed of sound using Eq. 2.69 is given by

$$a = \sqrt{1.4 \times 287 \times 288}\ \text{m/s}$$

With $R = 287\text{m}^2/\text{s}^2\text{K}$ and $\gamma = 1.4$ for air, we have $a = 340.17\ \text{m/s}$

The Mach number of the aircraft at sea level is

$$M = \frac{v}{a} = \left(\frac{1000}{3.6}\right)\left(\frac{1}{340.17}\right) = 0.817$$

At 11,000 m from ISA, temperature $T = -56.5\text{C}$; thus, $T = (273 - 56.5) = 216.5\ \text{K}$. The speed of sound by

$$a = \sqrt{1.4 \times 287 \times 216.5}\ \text{m/s}$$
$$a = 294.94\ \text{m/s}$$

The Mach number of the aircraft at 11,000 m altitude is $M = 0.942$. In addition, we can write

$$\frac{\Delta \rho}{\rho_{\mathrm{i}}} \approx \frac{1}{2}\left(\frac{V}{a}\right)^2 \approx \frac{M^2}{2}$$

Thus, the aircraft experiences different compressibility effects at the above two altitudes. The compressibility effects are particularly serious in this range (transonic range) of Mach numbers than any other range.

2.4.5 *Temperature Rise*

For a perfect gas as it was stated, we can write again

$$p = \rho RT \quad R = C_p - C_v \tag{2.70}$$

where C_p and C_v are specified heats at constant pressure and constant volume, respectively. Also, $\gamma = C_p/C_v$; therefore

$$R = \frac{\gamma - 1}{\gamma} C_p \tag{2.71}$$

For an isentropic change of state, an equation not involving T can be written as

$$\frac{p}{\rho^{\gamma}} = \text{constant} \tag{2.72}$$

Now, between state 1 and any other state, the relation between the pressures and densities can be written as

$$\left(\frac{p}{p_1}\right) = \left(\frac{\rho}{\rho_1}\right)^{\gamma} \tag{2.73}$$

Combining Eq. 2.73 and the equation of state, we get

$$\frac{T}{T_1} = \left(\frac{\rho}{\rho_1}\right)^{\gamma-1} = \left(\frac{p}{p_1}\right)^{\frac{\gamma-1}{\gamma}} \tag{2.74}$$

The above relations are very fruitful for gas dynamics, and they can be, expressed in terms of the Mach number. This relationship is an isentropic process under ideal gas condition [11].

Let us examine the flow around a symmetrical body, as shown in Fig. 2.8.

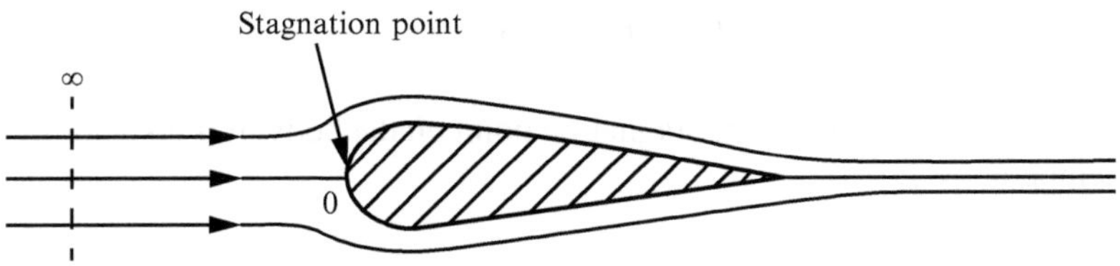

Fig. 2.8 Flow around a symmetrical body

In a compressible medium, there will be change in density and temperature at point 0. The temperature rise at the stagnation point can be obtained from the energy equation.

The energy equation for an isentropic flow is

$$h + \frac{v^2}{2} = \text{constant} \tag{2.75}$$

where h is the enthalpy.

Equating the energy at far, upstream ∞ and stagnation point 0, we get

$$h_\infty + \frac{V_\infty^2}{2} = h_0 + \frac{V_0^2}{2} \tag{2.76}$$

Since, at stagnation point 0 the velocity $V_0 = 0$,

$$h_0 - h_\infty = \frac{V_\infty^2}{2} \tag{2.77}$$

For a perfect gas, substituting integral result of Eq. 2.54, namely, $h = C_pT$ into Eq. 2.77 in above, we obtain

$$C_p(T_0 - T_\infty) = \frac{V_\infty^2}{2} \tag{2.78}$$

Combining Eqs. 2.69 and 2.71, results in the following relationship as

$$C_p = \frac{1}{\gamma - 1}\frac{a_\infty^2}{T_\infty} \tag{2.79}$$

Hence,

$$\Delta T = \frac{\gamma - 1}{2} T_\infty M_\infty^2 \tag{2.80}$$

For air, adiabatic index is $\gamma = 1.4$, and hence

$$T_0 = T_\infty \left(1 + 0.2M_\infty^2\right) \tag{2.81}$$

This is the temperature at the stagnation point on the body. It is also referred to as total temperature.

2.4.6 The Second Law of Thermodynamics

The second law stipulates that the total entropy of a system plus its environment cannot decrease; it can remain constant for a reversible process but must always increase for an irreversible process.

The first law of thermodynamics has been validated experimentally many times in many places. It is truly a law of physics. It always allows the conversion of energy from one form to another, but never allows energy to be produced or destroyed in the conversion process. However, it is not a complete description of thermal energy conversion processes. The first law would allow heat to be transferred from a cold body to a hot body as long as the amount of heat transferred decreased the internal energy of the cold body by the amount it increased the internal energy of the hot body.

However, this never happens. Heat can only be transferred from a hot body to a cold body. Therefore, there is a requirement for a law that explicitly states the direction of thermal energy transfer in addition to the conservation of energy expressed by the first law. This is the second law of thermodynamics. A simple statement of the second law would be that "heat cannot be spontaneously transferred from a cold body to a hot body." The second law of thermodynamics has been established around a simple cycle that is known as Carnot engine cycle and such simple engine in simple mode is illustrated in Fig. 2.9 [11].

In summary, the first law of thermodynamics provides the basic definition of internal energy, associated with all thermodynamic systems, and states the rule of conservation of energy. The second law is concerned with the direction of natural processes. It asserts that a natural process runs only in one sense and is not reversible. For example, heat always flows spontaneously from hotter to colder bodies, and never the reverse, unless external work is performed on the system. Its modern definition is in terms of entropy. The second law of thermodynamics states that the total entropy of an isolated system always increases over time or remains

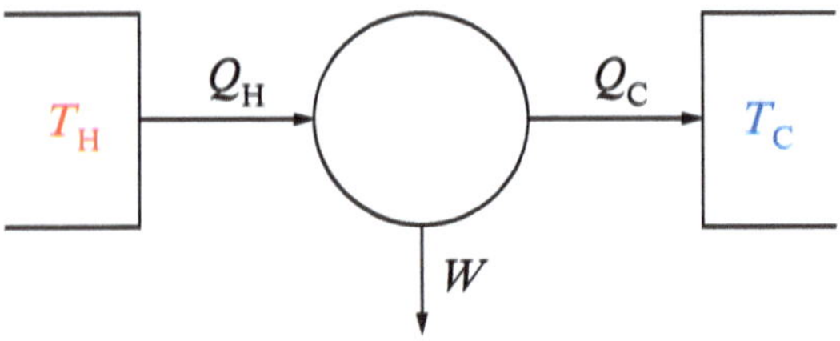

Fig. 2.9 The classical Carnot heat engine

constant in ideal cases where the system is in a steady state or undergoing a reversible process. The increase in entropy accounts for the irreversibility of natural processes and the asymmetry between future and past. The concept of entropy is defined in next step and denoted with the symbol S. Per famous German physicist Rudolf Clausius' (2 January 1822–24 August 1888) statement, the second law is as follows:

It is impossible to construct a device which operates on a cycle and whose sole effect is the transfer of heat from a cooler body to a hotter body.

According to his equality for a reversible process, we can write

$$\oint \frac{\delta Q}{T} = 0 \tag{2.82}$$

This means the line integral $\int_L \frac{\delta Q}{T}$ is path independent; thus, we can define a state function S, which we called it entropy, will satisfy the following:

$$dS = \frac{\delta Q}{T} \tag{2.83}$$

With this, we can only obtain the difference of entropy by integrating the above formula. The analysis result of Eq. 2.82 is presented in The Concept of Entropy below.

2.4.7 The Concept of Entropy

When the fundamental concepts of classical thermodynamics were being formulated the scientific community investigated a number of concepts that would formalize the second law analytically. The concept that was finally settled on was $\delta Q/T$. If a cyclic integral of this quantity is taken for the Carnot engine cycle in Fig. 2.9, it gives [11],

$$\oint \frac{\delta Q}{T} = \frac{Q_H}{T_H} - \frac{Q_L}{T_L} \tag{2.84}$$

Moreover, for the Carnot cycle,

$$\frac{Q_L}{Q_H} = \frac{T_L}{T_H} \quad \text{Or} \quad \frac{Q_H}{T_H} = \frac{Q_L}{T_L} \tag{2.85}$$

This then leads to the conclusion that

$$\oint \frac{\delta Q}{T} = 0 \tag{2.86}$$

As a result, $\delta Q/T$ is a perfect differential, and it can define a new property of a thermodynamic system. This property is called entropy and given the symbol S. Its differential is given by [11]

$$dS = \left.\frac{\delta Q}{T}\right|_{\text{reversible}} \tag{2.87}$$

In Eq. 2.87, $1/T$ serves as an integrating factor δQ. It can be integrated for a process in Fig. 2.10 to give the following condition as

$$\Delta S = \int_1^2 \frac{\delta Q}{T} \tag{2.88}$$

Hence, processing the integration of equation for full cycle in Fig. 2.10, we have

$$\begin{aligned}
\Delta S_{1-2} &= S_2 - S_1 = \int_1^2 \frac{\delta Q}{T} = \frac{Q_{H,1-2}}{T_H} \\
\Delta S_{2-3} &= S_3 - S_2 = \int_2^3 \frac{\delta Q}{T} = 0 \\
\Delta S_{3-4} &= S_4 - S_3 = \int_1^2 \frac{\delta Q}{T} = -\frac{Q_{L,3-4}}{T_L} \\
\Delta S_{4-1} &= S_1 - S_4 = \int_4^1 \frac{dQ}{T} = 0
\end{aligned} \tag{2.89}$$

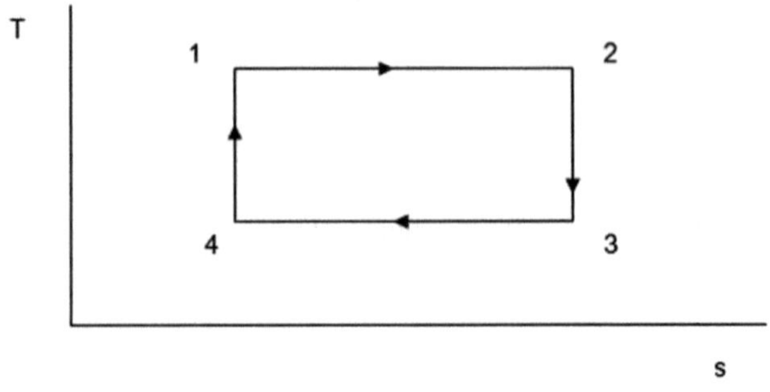

Fig. 2.10 Temperature entropy plot for a simple Carnot cycle in T-s diagram

For a reversible process, the heat transfer can be written as $dQ = TdS$. This allows the first law for closed systems to be written as

$$dU = \delta Q - \delta W = TdS - pdV \tag{2.90}$$

Additionally, for a unit, mass Eq. 2.90 will result in

$$du = Tds - pdv \tag{2.91}$$

The first law of thermodynamics for a flow system using the enthalpy per mass unit and usage of Eq. 2.91 gives

$$\begin{aligned} h &= u + pv \\ dh &= du + pdv + vdp \\ dh &= Tds + vdp \end{aligned} \tag{2.92}$$

Now consider an ideal gas with constant specific heats, and dividing Eq. 2.91 by temperature T as well as using the ideal gas law for unit mass results in [11]

$$ds = \frac{du}{T} + \frac{pdv}{T} = C_v \frac{dT}{T} + R\frac{dv}{v} \tag{2.93}$$

Equation 2.93 can be integrated to give

$$s_2 - s_1 = C_v \ln\left(\frac{T_2}{T_1}\right) + R\ln\left(\frac{v_2}{v_1}\right) \tag{2.94}$$

Equation 2.94 can also be evaluated as

$$\begin{aligned} ds &= C_p \frac{dT}{T} - \frac{v}{T}dp = C_p \frac{dT}{T} - R\frac{dp}{p} \\ s_2 - s_1 &= C_p \ln\left(\frac{T_2}{T_1}\right) + R\ln\left(\frac{p_2}{p_1}\right) \end{aligned} \tag{2.95}$$

Of course, these equations only apply to reversible processes. However, since they relate changes in entropy to other thermodynamic properties at the two end states, they can be used for reversible and irreversible processes [11].

If the entropy change is zero, called an isentropic process, then Eq. 2.94 becomes

$$\frac{T_2}{T_1} = \left(\frac{v_2}{v_1}\right)^{\gamma-1} \frac{T_2}{T_1} = \left(\frac{p_2}{p_1}\right)^{\frac{\gamma-1}{\gamma}} \Rightarrow \frac{p_2}{p_1} = \left(\frac{v_2}{v_1}\right)^{\gamma} \tag{2.96}$$

Of course, these equations are exactly the ones obtained earlier for a quasi-static adiabatic process [11].

2.4.8 Gas Dynamics Equations in Integral Form

Equations of gas dynamics can describe the fundamental laws of the mechanics of continuous medium, and one form that they may write in is are the integral form of these equations and they also may present both Eulerian and Lagrangian variables depending on the problem in hand.

In analyses and investigations of behavior of a compressible liquid or gas, one can regard the gas as a continuous medium, and various hydrodynamic (i.e., Lagrangian or Eulerian) models can describe the status of the motions of material in that medium [12].

For study of motion in compressible continuous one requires to have introductory to quantities such as vector like velocity $\vec{v}$, scalar, like temperature T and tensor, like pressure p, as well as others such as density ρ. This motion may take place in different coordinate systems, such as Eulerian, where the phenomenon of motion occurring at points in observer's frame of reference or this frame is fixed in space of laboratory. In this mode, we analyze the variation of the velocity, density, and other quantities associated with this motion at a given point. On the other hand, to follow the variation of the parameters of individual points of the medium, we take Lagrangian coordinate system, into consideration where frame of reference is in motion along the problem in hand.

In Euclidean space of three-dimensional coordinate presented with X_i, a certain coordinate system of frame, x_1, x_2, and x_3, that are fixed in space may be introduced to identify a coordinate of a point in this space with a position vector $\vec{r}$. The mathematical concept corresponding to physical notion of liquid flow then is that of a continuous transformation in this space into itself. In that, case the parameter time t is describing this transformation; in other words, we recognize concept of time if there is a motion involved. The value $t = 0$ can be taken at the initial instant, and the range of variation of t may be represented by the entire real axis.

Now if we are in Lagrangian frame of reference, we introduce a co-moving coordinate system of ξ_1, ξ_2, and ξ_3 and denote that by ξ_i for a Lagrangian point. Moreover, if we let a moving liquid particle to be located in a point $\xi_i = (\xi_1, \xi_2 \xi_3)$ at the instant $t = 0$ and in a point $\vec{r}(x_1, x_2, x_3) = x_1\hat{i} + x_2\hat{j} + x_3\hat{k}$ at the instant t, then, it can be assumed that $\vec{r}$ is defined as a function of ξ_i and time t, and from the kinematic standpoint, the flow of the medium can be regarded as the transformation as below [12]:

$$x_i = f(\xi_i, t) \quad i = 1, 2, 3 \tag{2.97}$$

If ξ_i is fixed and time t is varied, then Eq. 2.97 describes the trajectory of a particle M in space, located at the point ξ_i. If, on the hand, time t is fixed, then Eq. 2.97 defines a transformation of the region initially has been occupied by the liquid into the region occupied by the liquid at the instant t.

If initially different points remain different during the entire time of motion, we can introduce the inverse transformation as

$$\xi_i = \phi(\vec{r}, t) \tag{2.98}$$

Furthermore, we assume that both functions of f and ϕ are having continuous derivatives up to third order with respect to all variables in them, except for some singular surfaces, curves, or points.

In order to have the understanding of the fundamental laws of the mechanics of continuous medium, we need to utilize the essential equations of gas dynamics, and for that, we neglect effects due to disequilibrium, heat conduction, and viscosity for simplicity of driving these equations. The three most important gas dynamic equations are conservation of mass, momentum, and energy law, and their integral forms are briefly presented here. These integral forms are established for a certain liquid volume V^* individually as function of time t, bounded by a surface A. We leave the derivation of these integrals to the reader, and they can find them in any gas dynamics or transport phenomena textbook such as the one by Bird et al. [13]:

1. *Law of conservation of mass*

$$\frac{d}{dt}\int_{V^*(t)} \rho dV = 0 \tag{2.99}$$

2. *Law of conservation of momentum*

$$\frac{d}{dt}\int_{V^*(t)} \rho dV = \int_{V^*(t)} \vec{F} \rho dV - \oint_{A(t)} p\hat{n}\, dA \tag{2.100}$$

3. *Law of conservation of energy*

$$\frac{d}{dt}\int_{V^*(t)} \rho\left(e + \frac{v^2}{2}\right) dV = -\oint_{A(t)} p\vec{v}\hat{\boldsymbol{n}}\, dA + \int_{V^*(t)} \rho(\vec{F}\vec{v})\, dV \tag{2.101}$$

where

p = is the pressure
e = is the internal energy, which can be identified from enthalpy for an isentropic flow
$\vec{F}$ = is the mass force, which is the field vector of the external forces to a unit mass
$\hat{n}$ = is the unit vector pointed outward and normal to the surface A
v = velocity for unit mass in motion

If we want to account for heat condition with a heat influx, vector $\vec{q}$, then we apply the Fourier heat conduction law as

$$\vec{q} = -\kappa \Delta T \tag{2.102}$$

where κ is the thermal conductivity. Then it is necessary to add the integral form of Eq. 2.102, as it is shown below, to the right-hand side of Eq. 2.101:

$$-\oint_{A(t)} (\kappa \nabla T) \hat{\boldsymbol{n}}\, dA \tag{2.103}$$

Equations 2.99 through 2.101 hold for arbitrary volumes, particularly also, when volume V^* contains a surface A on which the hydrodynamic function becomes discontinuous.

In Euler variables, a continuous medium is described by the following fundamental quantities: [12]:

- Velocity vector $\vec{v}(\vec{r}, t)$
- Density $\rho(\vec{r}, t)$
- Pressure $p(\vec{r}, t)$

The internal energy is assumed as a rule to be specified by the thermodynamic function $e = e(p, \rho)$ or $e = e(T, \rho)$.

2.4.9 *Gas Dynamics Equations in Differential Form*

The continuous motion satisfies a system of differential equations that are the counterpart of the integral Eqs. 2.99 through 2.101. For gas motion in absence of mass forces $\vec{F}$ and dissipative processes, this system of differential equations is taking the following form:

1. *Mass conservation*

$$\frac{\partial \rho}{\partial t} + \nabla \cdot (\rho \vec{\mathbf{v}}) = 0 \tag{2.104}$$

2. *Momentum conservation*

$$\frac{\partial (\rho \vec{\mathbf{v}})}{\partial t} + \nabla \cdot \hat{P} = 0 \tag{2.105}$$

3. *Energy conservation*

$$\frac{\partial}{\partial t}\left(\rho e+\frac{\rho v^2}{2}\right)+\nabla\cdot\vec{\mathbf{v}}\left(\rho e+p+\frac{\rho v^2}{2}\right)=0 \tag{2.106}$$

where $e = e(p, \rho)$ and it depends on the properties of the gas and is assumed to be continuous and differentiable with respect to pressure p and density ρ.The symbol $\hat{P}$ is a tensor with components $P_{ij} = p\delta_{ik} + \rho v_i v_k$, while δ_{ij} is a unit tensor, and it is shown below:

$$\delta_{ik}=\begin{pmatrix}1 & 0\\ 0 & 1\end{pmatrix} \tag{2.107}$$

Furthermore, the function $e = e(p, \rho)$ for an ideal compressible medium such as an ideal gas has the following form:

$$e(p,\rho)=\frac{1}{\gamma+1}\frac{p}{\rho}+\text{constant} \tag{2.108}$$

The connection between the density, pressure, and temperature is given by thermal equation state of Eq. 2.70.

For adiabatic reversible gas motion, Eq. 2.106 can be transformed with the help of Eqs. 2.104 and 2.105 into a form as below:

$$\frac{de}{dt}-\frac{p}{\rho^2}\frac{d\rho}{dt}=0 \tag{2.109}$$

where $(d/dt) = (\partial/\partial t) + \nabla\cdot\vec{\mathbf{v}}$ is the substantive or total derivative that is also known as material derivative. More expansion of the system differential equations for gas dynamics can be found in any textbook [12, 13].

2.4.10 Perfect Gas Equation of State

At this stage, it is convenient to have some familiarity with concept of a perfect gas. From thermodynamics viewpoint, the perfect gas is the simplest working fluid, and it allows us to have better grasp of thermodynamic processes at granular level. From the aerospace point of view, the concept of perfect gas even becomes more important, since in this domain we deal with gases exclusively and often under near-perfect condition if not fully perfect gas per conceptual of definition such gas.

Thermal properties of gas measurement are indicating that for low densities, the thermal equation of state approaches the same form for all gases, namely,

$$Pv = R(\theta + \theta_0) \tag{2.110}$$

where θ_0 is the initial temperature, while θ is gas temperature at final stage and R is a characteristic constant for a particular gas and v is specific volume of gas. In terms of unite mass, where we have $\rho = 1/v$, the state equation 2.110 is written as

$$P = \rho R(\theta + \theta_0) \tag{2.111}$$

where θ_0 is a characteristic temperature which turns out to be the *same for all gases*, and hence it is useful to define an *ideal* or *perfect* gas which exactly satisfies Eq. 2.110. To be more precise, Eq. 2.110 defines a *family* of perfect gases, one for each value of R. Any gas at low enough density approaches to a perfect gas concept, with a particular value of R.

Since the characteristic θ_0 is found to be the same for all gases, one can define a new and more convenient temperature T, where $T = \theta + \theta_0$ and replacing this in Eq. 2.110 allows a new form of this equation as follows:

$$P = \rho RT \tag{2.112a}$$

In form of Eq. 2.112a temperature T is called as *absolute temperature*. It is also convenient to show that T, which is presenting the gas temperature here, has the same meaning for all thermodynamic systems. The scale and the zero point of T are determined from the scale and zero point of the thermometer used to measure θ_0. Thus, in the centigrade scale, one finds that $\theta_0 = 273.16°$, and in the Fahrenheit, $\theta_0 = 459.69°$. Thus, the absolute temperature T can be written as

$$\begin{aligned} T &= \theta + 273.16 \quad \text{Degrees Centigrade absolute or Kelvin (°K)} \\ T &= \theta + 459.69 \quad \text{Degrees Fahrenheit absolute or Rankine (°R)} \end{aligned}$$

The constant characteristic R in Eq. 2.112a has the dimensions of (velocity)^2/temperature, and it is related to the velocity of sound a in the gas as $R \approx (da^2/dT)$. If we reform Eq. 2.112a for a given mass M by putting $\rho = M/V$, we have the following:

$$PV = MRT \tag{2.112b}$$

A study of the behavior of different gases led very early to the concept that gases are composed of molecules and that the characteristic parameter of the family of perfect gases defined by Eqs. 2.112a and 2.112b is the mass of these molecules. Thus, Eq. 2.112a can be written in terms of a dimensionless mass ratio $\mu = M/m$, where m denotes the mass of one gas molecule. Written in this reduced or "similarity" form, the family of Eqs. 2.112a, 2.112b, 2.112c and 2.112d can be reduced to a single one as

$$\frac{PV}{\mu} = kT \tag{2.112c}$$

where k is a *universal* constant, which is known to us as Boltzmann constant. Instead of m, the mass of a molecule, one often uses the "molecular weight" m in relative units such that $\mathfrak{M}_{\text{Oxygen}} = 32$. In terms of $\mathfrak{M}$, one has:

$$\frac{PV}{\mu} = \mathfrak{R}T \tag{2.112d}$$

where $\mu = M/m$ and $\mathfrak{R}$ is called the *universal gas constant.* We can also use as unit mass the *mole* and thus make $\mu = 1$ in Eq. 2.112d. Then, V becomes the *mole volume*. In this book, we shall not use the mole, but we shall continue to refer to unit mass where it is required and for sure uses the term perfect gas in the form of Eq. 2.112a.

It is worth to state that the internal energy of a perfect gas U is a function of temperature only and written it as

$$U = U(T) \tag{2.113}$$

Equation 2.113 can be, taken as a result of experience and experiment and often a gas called "calorically perfect," if Eq. 2.113 simplifies further to the following form:

$$U = \text{constant} \cdot T \tag{2.114}$$

Equation 2.114 does not obey directly from Eq. 2.112a by pure thermodynamic reasoning; however, for a certain range of temperature, it can be justified by experience, and it also follows the rules of statistical mechanics.

2.5 Unsteady Motion of Continuous Media and Self-Similarity Methods

In modern physics, the sciences of methods of the mechanics of continuous media are assuming an ever-increasing importance. Field of gas dynamics covers the flow of all compressible media and covers liquids and in case of solids under conditions of high pressure. The study of such flow involves not only the methods of mechanics but also methods of other branches of physics, particularly thermodynamics. In order to have a concept of an unsteady motion, we need to understand the parameters characterizing the state of motion of the medium. If such parameters and their characteristics are invariant with time in the region where the motion of medium is studied, then this motion is *steady* (stationary) and if these parameters change with time (variant), then motion is called *unsteady*.

Study of *unsteady motion* has various applications, and solution of such motion raises huge interest in many applied technical problems, such as understanding of the motion of explosion and implosion products of the medium in which these events take place, and a few examples were mentioned in the first few chapters of this book. The investigation of fluctuations of gas inside various engineering and fundamental problems of modern physics in the field of cosmology is another example of such solution. In fact, various processes occurring in the universe, such as the formation of stars or the gigantic eruptions that originate in the sun and stars or understanding of laser-driven fusion (National Ignition Facility) program within recent study of fusion energy, are caused by unsteady motion of tremendous masses of gas.

The development of gas dynamics is the result of work by many scientists of many countries. These developments are starting from the basic equations of gas dynamics that are derived from three fundamental laws of nature, namely:

1. Conservation of mass
2. Conservation of momentum
3. Conservation of energy

To study the motion of medium, we need to determine the three components of velocity, the density, and the pressure of the medium as function of three spatial coordinates and time. The components of the vector of conservation of mass and the law of conservation of energy give two equations; thus, to determine the five unknown functions, we have a system of five equations, which they represent partial differential equations of the first order. These equations are shown here as part of the behavior of a lossless one-dimensional fluid described by the following set of conservation equations, also known as *Euler's equations*:

$$\frac{\partial \rho}{\partial t} + \frac{\partial(\rho v)}{\partial x} = 0 \quad \text{Conservation of Mass} \tag{2.115a}$$

$$\frac{\partial(\rho v)}{\partial t} + \frac{\partial(\rho v^2 + p)}{\partial x} = 0 \quad \text{Conservation of Momentum} \tag{2.115b}$$

$$\frac{\partial(\rho e)}{\partial t} + \frac{\partial(\rho v e + p v)}{\partial x} = 0 \quad \text{Conservation of Energy} \tag{2.115c}$$

where $\rho = \rho(x, t)$ is density, $v = v(x, t)$ is volume velocity, $p = p(x, t)$ is absolute pressure, and $e = e(x, t)$ is total energy, which is internal plus kinetic all together. The three equations are not complete without a constitutive relation among the four dependent variables. More people, in their treatment of hydrodynamics, often leave out the energy equation and make an assumption of the type $p = G(\rho)$, which essentially reduces the system to a two-variable system (in ρ and υ). For gas dynamics, we assume polytropic gas behavior

$$e = \frac{\upsilon^2}{2} + \frac{p}{\rho(\gamma + 1)} \tag{2.116}$$

where $\gamma > 1$ (i.e., also known as adiabatic index) is a constant, which follows directly from thermodynamics as well as our previous discussion earlier, and it is equal to the ratio of specific heats $\gamma = C_p/C_v$, where C_p and C_v are heat capacity at constant pressure and volume, respectively.

Note: The dependent variables in order to optimize the stability condition on the resulting system is scaled as

$$\hat{\upsilon} = \frac{\upsilon}{\upsilon_0} \quad \hat{p} = \frac{p}{p_0} \quad \hat{\rho} = \frac{\rho \upsilon_0^2}{p_0} \tag{2.117}$$

The parameters υ_0 and p_0 have dimensions of velocity and pressure, respectively, and they are nondimensionalized system. υ_0 will again become the space–step/time–step ratio in the numerical simulation routine and plays a role similar to that of r_0 and follows directly from physical considerations.

Using the energy density definition (Eq. 2.116), the system (Eqs. 2.115a, 2.115b and 2.115c) can be written in nonconservative form (after some tedious algebraic manipulations) as

$$\begin{bmatrix} 1 & 0 & 0 \\ 0 & \hat{\rho} & 0 \\ 0 & 0 & 1 \end{bmatrix} \frac{\partial}{\partial t'} \begin{bmatrix} \hat{\rho} \\ \hat{\upsilon} \\ \hat{p} \end{bmatrix} + \begin{bmatrix} \hat{\upsilon} & \hat{\rho} & 0 \\ 0 & \hat{\rho}\hat{\upsilon} & 1 \\ 0 & \gamma\hat{p} & \hat{\upsilon} \end{bmatrix} \frac{\partial}{\partial x} \begin{bmatrix} \hat{\rho} \\ \hat{\upsilon} \\ \hat{p} \end{bmatrix} = 0 \tag{2.118}$$

The equations of gas dynamics can be written in two different forms, and they are represented as follows:

1. *Eulerian form*
2. *Lagrangian form*

These two above forms are discussed in the next two subsections where we just present the conservation equation without any proof and their derivations are left to our readers or they can refer to the book by Stanyukovich [8] or Bird et al. [13].

In the most general case of three-dimensional spatial motion of a medium, the law of conservation and the equations of state enable us to derive *six fundamental equations* of hydrodynamics or gas dynamics:

- Three equations of motion, one for each dimension of x_1, x_2, and x_3 that obey the law of conservation of momentum
- One equation each from the law of conservation of mass
- One equation for the law of conservation of energy that obeys this law

- One equation for equation of state of the medium, in order to complete the set of six equations

These six fundamental equations determine the six unknown quantities that characterize the motion and state of the medium: three components of velocity, density, pressure, and entropy or temperature of the medium [8].

The main use of thermodynamic methods in hydrodynamics or gas dynamics is to connect by means of thermodynamic equations the basic parameters of the state in medium, namely, density, pressure, temperature, entropy, and the heat content as well as the velocity of sound. In case of an ideal gas, this problem becomes trivial. However, in case of dense media, the problem is much more difficult to deal with, since the equation of state is more complicated than in an ideal gas. Hence, a suitable approximation and utilization of asymptotic method to the equation of state and to the isentropic equation make it possible to solve completely many problems in the motion of solid media [8].

Furthermore, with help of self-similarity approach, we can study the spatial motion of the medium via either Eulerian or Lagrangian schema.

2.5.1 Fundamental Equations of Gas Dynamics in the Eulerian Form

In this form, it is possible in principle to determine at each given instant of time the distribution of the six fundamental parameters as functions of any three spatial coordinates as we know distribution of these three parameters in space that is equivalent to determining at each specified point in space. The variations of the above six parameters with time and their time dependency will be measured against a fixed frame of reference.

Here, we are just presenting the fundamental equations of gas dynamics from conservation viewpoint and equation state using the Cartesian, cylindrical and spherical coordinate systems in the Eulerian form.

1. *Cartesian coordinate* (x, y, z)

$$
\begin{aligned}
&\frac{\partial\rho}{\partial t}+\frac{\partial}{\partial x}(\rho u)+\frac{\partial}{\partial y}(\rho v)+\frac{\partial}{\partial z}(\rho w)=0 && \text{Equation of Continuity or Momentum}\\
&\left.\begin{aligned}
&\frac{\partial u}{\partial t}+u\frac{\partial u}{\partial x}+v\frac{\partial u}{\partial y}+w\frac{\partial u}{\partial z}+\frac{1}{\rho}\frac{\partial p}{\partial x}=0\\
&\frac{\partial v}{\partial t}+u\frac{\partial v}{\partial x}+v\frac{\partial v}{\partial y}+w\frac{\partial v}{\partial z}+\frac{1}{\rho}\frac{\partial p}{\partial y}=0\\
&\frac{\partial w}{\partial t}+u\frac{\partial w}{\partial x}+v\frac{\partial w}{\partial y}+w\frac{\partial w}{\partial z}+\frac{1}{\rho}\frac{\partial p}{\partial z}=0
\end{aligned}\right] && \text{Equation of Motion}\\
&\frac{\partial\rho}{\partial t}+u\frac{\partial e}{\partial x}+v\frac{\partial e}{\partial y}+w\frac{\partial e}{\partial z}=0 && \text{Equation of Eenergy}\\
&\mathrm{e}=\mathrm{e}(p,\rho) && \text{Energy}\\
&p=p(\rho,e) && \textit{Pressure}\\
&\rho=\rho(\vec{\mathbf{v}},t) && \text{Density}\\
&\vec{\mathbf{v}}=(u,v,w) && \text{Velocity}
\end{aligned}
\tag{2.119}
$$

2. *Cylindrical coordinate* (x,θ,z)

$$
\begin{aligned}
&\frac{\partial\rho}{\partial t}+\frac{1}{r}\frac{\partial}{\partial r}(\rho r u)+\frac{1}{r}\frac{\partial}{\partial\theta}(\rho v)+\frac{\partial}{\partial z}(\rho w)=0 && \text{Equation of Continuity or Momentum}\\
&\left.\begin{aligned}
&\frac{\partial u}{\partial t}+u\frac{\partial u}{\partial r}+\frac{v}{r}\frac{\partial u}{\partial\theta}+w\frac{\partial u}{\partial z}-\frac{v^2}{r}+\frac{1}{\rho}\frac{\partial p}{\partial r}=0\\
&\frac{\partial v}{\partial t}+u\frac{\partial v}{\partial r}+\frac{v}{r}\frac{\partial v}{\partial\theta}+w\frac{\partial v}{\partial z}+\frac{uv}{r}+\frac{1}{\rho r}\frac{\partial p}{\partial\theta}=0\\
&\frac{\partial w}{\partial t}+u\frac{\partial w}{\partial r}+\frac{v}{r}\frac{\partial w}{\partial\theta}+w\frac{\partial w}{\partial z}+\frac{1}{\rho}\frac{\partial p}{\partial z}=0
\end{aligned}\right] && \text{Equation of Motion}\\
&\frac{\partial\rho}{\partial t}+u\frac{\partial e}{\partial r}+\frac{v}{r}\frac{\partial e}{\partial\theta}+w\frac{\partial e}{\partial z}=0 && \text{Equation of Eenergy}\\
&\mathrm{e}=\mathrm{e}(p,\rho) && \text{Energy}\\
&p=p(\rho,e) && \textit{Pressure}\\
&\rho=\rho(\vec{\mathbf{v}},t) && \text{Density}\\
&\vec{\mathbf{v}}=(u,v,w) && \text{Velocity}
\end{aligned}
\tag{2.120}
$$

3. *Spherical coordinate* (x, θ, φ)

$$\frac{\partial \rho}{\partial t}+\frac{1}{r^2}\frac{\partial}{\partial r}(\rho r^2 u)+\frac{1}{r\sin\theta}\frac{\partial}{\partial\theta}(\rho\upsilon\sin\theta)+\frac{\partial}{r\sin\theta}(\rho w)=0 \qquad \text{Equation of Continuity or Momentum}$$

$$\left.\begin{array}{l}
\dfrac{\partial u}{\partial t}+u\dfrac{\partial u}{\partial r}+\dfrac{\upsilon}{r\sin\theta}\dfrac{\partial u}{\partial\theta}+\dfrac{w}{r}\dfrac{\partial u}{\partial\theta}-\dfrac{\upsilon^2+w^2}{r}+\dfrac{1}{\rho}\dfrac{\partial p}{\partial r}=0\\[2ex]
\dfrac{\partial \upsilon}{\partial t}+u\dfrac{\partial \upsilon}{\partial r}+\dfrac{\upsilon}{r\sin\theta}\dfrac{\partial \upsilon}{\partial\varphi}+\dfrac{w}{r}\dfrac{\partial \upsilon}{\partial\theta}+\dfrac{u\upsilon}{r}-\dfrac{\upsilon w\cot\theta}{r}+\dfrac{1}{\rho r\sin\theta}\dfrac{\partial p}{\partial\theta}=0\\[2ex]
\dfrac{\partial w}{\partial t}+u\dfrac{\partial w}{\partial r}+\dfrac{\upsilon}{r\sin\theta}\dfrac{\partial w}{\partial\varphi}+\dfrac{w}{r}\dfrac{\partial w}{\partial\theta}\dfrac{uw}{r}-\dfrac{\upsilon^2\cot\theta}{r}+\dfrac{1}{\rho r}\dfrac{\partial p}{\partial\theta}=0
\end{array}\right] \qquad \text{Equation of Motion}$$

$$\frac{\partial \rho}{\partial t}+u\frac{\partial e}{\partial r}+\frac{\upsilon}{r}\frac{\partial e}{\partial\theta}+w\frac{\partial e}{\partial z}=0 \qquad \text{Equation of Eenergy}$$

$$\mathrm{e}=\mathrm{e}(p,\rho) \qquad \text{Energy}$$

$$p=p(\rho,e) \qquad \textit{Pressure}$$

$$\rho=\rho(\vec{\mathbf{v}},t) \qquad \text{Density}$$

$$\vec{\mathbf{v}}=(u,\upsilon,w) \qquad \text{Velocity}$$

$$(2.121)$$

2.5.2 *Fundamental Equations of Gas Dynamics in the Lagrangian Form*

The equations of gas dynamics, written in Lagrangian form, describe the motion of each individual particle. Solution of these equations determine the coordinates and the variables of state of this particle at any instant or time t, starting with a certain arbitrary chosen initial time t_0. In the absence of external forces, the Lagrangian form of the equations of motion in a Cartesian coordinate system is given as follows, and it is specified at the stating of motion $t=0$:

$$\begin{cases}
\dfrac{\partial u}{\partial t}=\dfrac{\partial^2 x}{\partial t^2}=-\dfrac{1}{\rho}\dfrac{\partial p}{\partial x}\\[2ex]
\dfrac{\partial \upsilon}{\partial t}=\dfrac{\partial^2 y}{\partial t^2}=-\dfrac{1}{\rho}\dfrac{\partial p}{\partial y}\\[2ex]
\dfrac{\partial w}{\partial t}=\dfrac{\partial^2 z}{\partial t^2}=-\dfrac{1}{\rho}\dfrac{\partial p}{\partial z}
\end{cases} \qquad \text{Equation of Motion} \qquad (2.122)$$

Note that the specified values of the initial coordinates of any particle in rectangular coordinate system is $x_0=a$, $y_0=b$, and $z_0=c$, so that the Equation 2.122 is valid for the current coordinates of the particle at x, y, and z and become function of the time t and of the values of the coordinates a, b, c as well.

Derivation of continuity equation as well as energy equation in Lagrangian system both are provided by Stanyukovich [8], and we just write the results here:

$$\rho\frac{d\Delta}{dt}+\frac{d\rho}{dt}\Delta=0$$
$$\frac{d\Delta}{dt}=-\frac{1}{\rho}\frac{d\varphi}{dt}\Delta=-\frac{1}{\rho}\Delta\left(\frac{\partial\rho}{\partial t}+\frac{\partial\rho}{\partial x}u+\frac{\partial\rho}{\partial y}v+\frac{\partial\rho}{\partial z}w\right)\equiv\Delta\text{div}\,\vec{v}\quad \text{Equation of Momentum} \tag{2.123}$$

and

$$\frac{de}{dt}=\left(\frac{\partial e}{\partial t}\right)_{a,b,c}=0 \quad \text{Equation of Energy} \tag{2.124}$$

From which it follows that $e=e(a,b,c)$.

For other coordinate systems, such as cylindrical and spherical, the detailed derivation of the equations can be found in [8].

2.6 Study of Shock Waves and Normal Shock Waves

One of the important phenomena that could be studied by dimensional analyses and consequently utilization of self-similarity method/approach is the problem of shock waves both continuous and discontinuous transitions between subsonic and supersonic or vice versa in a medium such as fluid or gas flow.

Continuous transition from supersonic to subsonic flow or vice versa is taking place at a nuzzle throat or duct, and this problem is very well known, and various solutions are offered by different scientists and researchers in their published articles or textbooks (i.e., Shapiro). Shock waves are created by a discontinuous transition between two points in the flow, which we mean a sudden change in the velocity, pressure, density, etc.

We can easily discover that only discontinuous transition between two conditions of subsonic and supersonic is a source of shock waves. A complete study of this subject is beyond the scope of this book, and we retain certain aspects of the subject shock waves that we need to encounter in this book, yet we recommend the reader to refer her- or himself to the most amiable classical books out there.

The term normal shock is used to express the fact that shock is perpendicular to the flow direction, before and after transition through the shock wave. The above expression implies that there is no change in flow direction as a result of passing through shock wave. Shocks generated this way also are as a result of compression waves that are seen in nozzles, turbomachinery blade passages, etc.

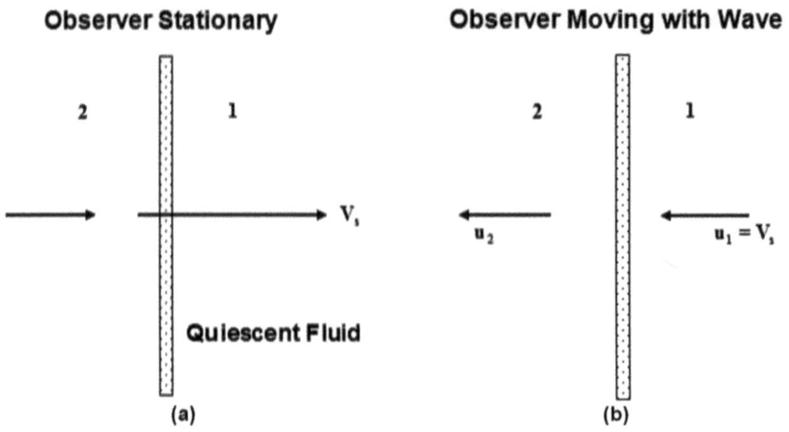

Fig. 2.11 Illustration of normal shock wave propagating in the air

The compression process across the shock wave is highly irreversible, and so it is undesirable in such case.

Illustration of normal shock-wave propagation through a quiescent medium, i.e., air, is depicted in Fig. 2.11, where the symbol of V_s is shock speed in the laboratory frame of reference and u_1 as well as u_2 are the speed of shock propagation in point (1) and (2) across the wave. Assuming that the medium is air that shock is propagating, then at point (1), an acoustic wave travels with the speed of sound, while at point (2), the changes in properties across an acoustic wave are infinitesimal and isentropic, and they are large and irreversible across a normal shock wave.

The governing conservation equations in one-dimensional frame for frictionless, adiabatic, steady flow of a calorically perfect gas are written as

$$\rho_1 u_1 = \rho_2 u_2 \tag{2.125}$$

$$p_1 + \rho_1 u_1^2 = p_2 + \rho_2 u_2^2 \tag{2.126}$$

$$h_1 + \frac{1}{2}u_1^2 = h_2 + \frac{1}{2}u_2^2 \tag{2.127}$$

$$\begin{aligned} s_2 - s_1 &= C_v \ln\frac{T_2}{T_1} + R \ln\frac{v_2}{v_1} \\ &= C_v \ln\frac{T_2}{T_1} + C_p \ln\frac{v_2}{v_1} \\ &= C_p \ln\frac{T_2}{T_1} - R \ln\frac{p_2}{p_1} \end{aligned} \tag{2.128}$$

It can be seen from the energy equation that the stagnation temperature is constant across the shock wave, as there is no heat addition or removal.

The mathematical derivation of the normal shock-wave solution can be found in any fundamental gas dynamics classical textbook such as the one by Babu [9].

Rankine–Hugoniot shock wave also can be derived as a result of mathematical derivation of normal shock, and they are written here as

$$\frac{p_2}{p_1} = \frac{\left[\frac{v_2}{v_1} - \frac{\gamma+1}{\gamma-1}\right]}{\left[1 - \frac{v_2}{v_1}\frac{\gamma+1}{\gamma-1}\right]} \tag{2.129}$$

All the variables and quantities in above equations are defined as before.

2.6.1 Shock Diffraction and Reflection Processes

There are two phenomena, which can result from interaction of a shock wave with a solid surface. A shock wave reflects off a surface if the angle of incidence is less than 90°. If the angle of incidence more than 90°, then the shock wave diffracts over a surface. However, it is necessary to note that the diffraction phenomenon does not exist in steady flows. Moreover, for steady flows, the term head-on reflection of a shock wave loses any meaning.

In contrast to steady flows, the interaction (i.e., reflection or diffraction) of a moving shock wave with a solid surface is the process developing in time. Reflection and diffraction of a planar shock wave, having constant parameters behind the fronts, over straight wedges are the pseudo-stationary processes. In this case, we need to tell about reflection or diffraction of a single shock wave.

The key to understand the processes of reflection and diffraction of a single shock wave in a pseudo-steady flow falls in the idea that the whole process should be considered as a combination of two subprocesses:

1. An incident shock wave front reflection or diffraction
2. An incident shock-induced flow deflection around the leading edge of a ramp

In the case of single shock diffraction, a complex inner structure containing a separation zone, a vortex, and a stagnation shock-wave configuration is formed. In case of shock-wave reflection, we observe a variety of reflection types.

The reflection and diffraction of a single shock wave in pseudo-steady flow are the most investigated phenomena, and results can be found in an open public domain.

References

1. L.I. Sedov, *Mechanics of Continuous Media*, vol I and II, 4th edn. (World Scientific Publishing, Singapore, 1997)

2. F.H. Harlow, A machine calculation method for hydrodynamic problems. Los Alamos Scientific Laboratory report LAMS-1956 (1955)
3. J.M. Dawson, Particle simulation of plasmas. Rev. Mod. Phys. **55**(2), 403 (1983)
4. G.I. Taylor, The formation of a blast wave by a very intense explosion, I. Theoretical discussion. Proc. Roy Soc. A **201**, 159–174 (1950)
5. G.I. Taylor, The formation of a blast wave by a very intense explosion, II. The atomic explosion of 1945. Proc. Roy Soc. A **201**, 175–186 (1950)
6. G. Guderley, Starke kugelige und zylindrische Verdichtungsstosse in der Nahe des Kugelmittelpunktes bzw. der Zylinderachse. Luftfahrt-Forsch **19**, 302–312 (1942)
7. B. Zohuri, *Dimensional Analysis and Self-Similarity Methods for Engineers and Scientists*, 1st edn. (Springer, Cham, 2015)
8. K. Stanyukovich, *Unsteady Motion of Continuous Media* (Pergammon Press, New York, 1970)
9. V. Babu, *Fundamentals of Gas Dynamics*, 2nd edn. (Wiley, Chichester, 2015)
10. E. Rathakrishnan, *Gas Dynamics* (Prentice-Hall of India, New Delhi, 2004)
11. B. Zohuri, P. McDaniel, *Thermodynamics In Nuclear Power Plant Systems*, 1st edn. (Springer, Cham, 2015)
12. V.P. Korobeinikov, *Problem of Point Blast Theory* (American Institute of Physics, New York, 1991)
13. R.B. Bird, W.E. Stewart, E.N. Lightfoot, *Transport Phenomena*, 2nd edn. (Wiley, New York, 2001)

Chapter 3
Shock Wave and High-Pressure Phenomena

The propagation of self-sustained gaseous detonations either implosion or explosion is a complex, multidimensional process involving interactions between incident shocks, Mach stems, transverse waves, and boundaries of the regions through which the detonation is moving. In this chapter, we are interested in problems that fall into one-dimensional process categories in particular when they are involved with an implosion or explosion of homogeneous and symmetric types. Self-similarity offers an excellent simplified solution, by reducing complex sets of equation to some simple ordinary sets of differential equations, where a simple exact solution can be found. Here we study well-known problems of implosion and explosion of symmetry nature, where a three-dimensional problem has reduced to one-dimensional status and obeying either Lagrangian or Eulerian schema or in some cases the problem has followed an Arbitrary Lagrangian–Eulerian (ALE) roles.

3.1 Introduction to Blast Waves and Shock Waves

The blast wave is a description of an event that is an instantaneous release of energy at a point in real air or in case of ideal gas for the purpose of analyzing the physics of point blast theory and shock waves, generated as a result of it. By definition, strong pressure wave in any elastic medium generates shockwaves and these elastic medium could be air, water, or a solid substance, and in that case, sources of generation may be a supersonic object such as jet aircraft, source of explosions, or other phenomena that cause violent changes in pressure of that media.

Shock waves differ from sound waves in that the wave front in which compression takes place is a region of sudden and violent changes in parameters of medium, such as stress, density, and temperature. Therefore, because of these sudden changes, shock waves propagate in a manner different from that of ordinary acoustic waves. Furthermore, shock waves travel faster than sound, and their

B. Zohuri, *Dimensional Analysis Beyond the Pi Theorem*,
DOI 10.1007/978-3-319-45726-0_3

speed increases as the amplitude is raised, but the intensity of a shock wave also decreases faster than does that of a sound wave, because some of the energy of the shock wave is expended to heat the medium in which it travels.

The amplitude of a strong shock wave, as created in air by an explosion, decreases almost as the inverse square of the distance until the wave has become so weak that it obeys the laws of acoustic waves. Shock waves alter the mechanical, electrical, and thermal properties of solids and, thus, can be used to study the equation of state (a relation between pressure, temperature, and volume) of any material.

On the other hand, point blast theory needs to obey a need for describing phenomena produced in continuous media by the explosion that takes place via strong blast of small size and weight of element with high specific energy. Thus, the study of physics of blast waves generated by such source becomes our concern. In case of point blast theory, it deals with phenomena such as strong explosions in the atmosphere, where explosive processes such as propagation of shock waves become a nature of this theory to study. Similarly, we can observe this type of theory in physics of nuclear or thermonuclear detonations. Other examples of encountering the blast wave and resulted shock waves are in inertial fusion confinement, where laser or a heavy ion beam drives a tiny micro-balloon fueled by two isotopes of hydrogen, namely, deuterium (D) and tritium (T), to ignition temperature, in order for the fusion to take place, between these two elements.

In that case, we are looking at the shock waves created by two aspects of this process, namely, implosion pushing fuel toward the corona of pellet, approaching fusion temperature and explosion, at the ablation surface of pellet that generates outward shockwaves. In any case, we let a small explosive mass be concentrated in a volume much smaller than the ambient medium and let an explosion be created at some instant of time, and it will release energy of per unit volume much higher than in ambient medium. Therefore, when it comes to explosion or implosion phenomena, the principle idealization is the assumption that the energy is released instantaneously, while the volume occupied by the explosive and the mass of the charge are zero.

3.2 Self-Similarity and Sedov–Taylor Problem

The mathematical formulation of the problem of the nuclear explosion and the estimation of its mechanical and physical effects on the surroundings were itself a challenging task. There was hardly any literature on this subject. Therefore, some of the best minds in applied mathematics and physics were made to put their heads together to unravel this topic. This gave a great fillip to nonlinear science, which has since made great strides and which now permeates and influences all sciences—pure and applied.

The explosion problem in a perfect gas could be considered for the case when initial velocity, density, and pressure is, assumed to be, uniform. Many authors have studied the motion of diverging spherical and cylindrical shock waves in a perfect gas, for a homogeneous and symmetrical form [1–6].

The diverging spherical shock waves for Trinity explosion of fission atomic bomb were studied by Taylor (1950a) [1] and Sedov (1969) [7]. There are few examples that are mentioned here, although there were other authors that independently did a similar study, and this is the class of solutions known as self-similar solutions of the first kind. Taylor (1950a) [1] demonstrated the existence of the self-similar solutions for a shock wave propagating in the vicinity of the center of divergence.

Mathematically, the continuous flow behind the shock is governed by the non-isentropic equations of gas dynamics, which must be solved subject to the so-called Rankine–Hugoniot conditions at the shock and the symmetry condition at the center requiring that the particle velocity there is zero. Along the shock trajectory, the theory of shocks imposes more boundary conditions than are appropriate to the given system. This overdetermined data, however, leads to the finding of the shock trajectory, which itself is an unknown factor. This, in this sense, constitutes a free boundary value problem. In this simplest model, the role of heat conduction is ignored. Taylor (1950) [1] made some highly intuitive physical statements about this phenomenon [2].

For example, he observed that the explosion forces most of the air within the shock front into a thin shell just inside the front (see Figs. 3.1 and 1.8a, b as well). This is the subject of discussion for this section and forms the basis of an analytic theory of blast waves in an exponential atmosphere by Laumbach and Probestein (1969) [3]. Taylor (1950) [2] also observed that as the front expands, the maximum pressure decreases till at about 10 atm with the analysis under the assumption of an infinitely strong shock ceases to hold.

On July 16, 1945, the first atomic bomb ever was detonated in New Mexico. The pictures shown in Figs. 3.1 and 1.8a, b were released and published in Life Magazine. The energy of the blast, however, was highly classified and it was kept secret. The story goes that Geoffrey Ingram Taylor, the British physicist, used dimensional analysis to estimate the latter energy from the data available in the pictures. The analysis was presented in Chap. 1 of the book, the purpose of present mentioning of the same problem is to guide you through Taylor's analysis, and formulation of the similarity solution was derived entirely from physical arguments.

What is seen in the pictures is a spherical shock wave separating the undisturbed air from the region affected by the explosion. As usual, a dimensional analysis is simplified by some educated guess. Taylor's analysis is based on the following assumptions:

- The explosion itself is so rapid that the only relevant characteristic of the bomb is the amount of energy E that is releases. The duration of the explosion is irrelevant.
- The shock wave propagation is so quick that it can be modeled as an adiabatic process, characterized by adiabatic exponent (i.e., adiabatic index) γ.

Fig. 3.1 Trinity nuclear test, New Mexico on July 16, 1945

- The pressure generated by the shock is much larger than the atmospheric pressure so that the latter should not be accounted for in the analysis. Only the density of the air ρ_0 matters.

Based on these simplifying assumptions, use dimensional analysis to find the way in which the radius R of the shock wave increases, with time t, and it can be seen as a function of $R = f(t, E, \rho, \gamma)$, and this function can be established from classical mechanics theory of kinetic energy. By using dimensional arguments, he wrote the similarity form of the solution in Eulerian coordinates in terms of the similarity variables r/R, where R, the radius of the shock, was found to be proportional to $t^{2/5}$; he did not use any sophisticated transformation theory of nonlinear partial differential equations (PDEs). Taylor reduced the system of nonlinear PDEs to nonlinear ordinary differential equations (ODEs) and numerically solved the latter, subject to the strong shock conditions (appropriately transformed), and the requirement of spherical symmetry, namely, that the particle velocity at the center of the explosion, must be zero. He also used the conservation of total energy, E, behind the shock to derive the shock trajectory. The constant $B = E/\rho_0 A^2$, which appears in the shock law $R = Bt^{2/5}$, involves the nondimensional form of energy and was found from the numerical solution; it varies with adiabatic index γ the ratio of specific heats as $\gamma = C_p/C_v$, where C_p and C_v are specific heat at constant pressure and volume, respectively.

The general solution of the Chap. 1 of the book is as follows:

$$f(\gamma) = \frac{\rho R^5}{Et^2}. \tag{3.1}$$

Equation 3.1 above predicts a propagation of the shock according to a $R \simeq t^{2/5}$ power law, as it is stated before. This law is extremely well, followed by the data as it can be seen in Fig. 3.2 here.

The general solution of Taylor's problem in term of energy releases from fission nuclear explosion is then given as follows:

$$E = \left(R^5 \rho_0\right)/t^2. \tag{3.2}$$

Table 3.1 here presents the values of R as a function of t, determined from the released pictures of the Trinity test explosion in 1947.

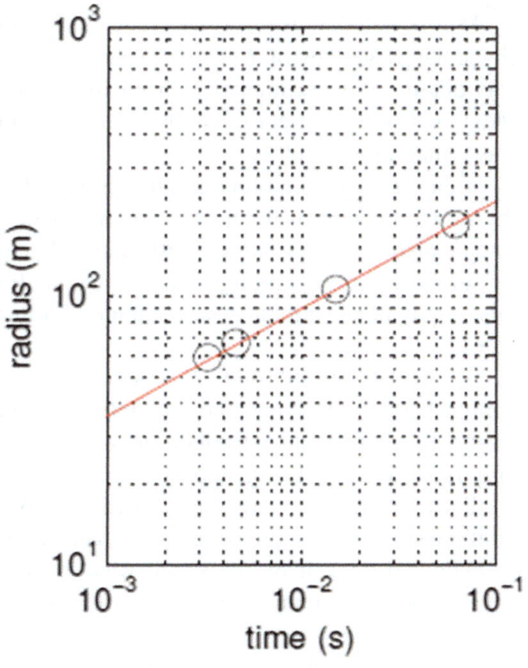

Fig. 3.2 Comparison of the data from the released Trinity test pictures (*small circles*) and power law (*red line*)

Table 3.1 Time-dependent radius of the Trinity shock wave, as determined from the released pictures

Time (ms)	3.3	4.6	16	62
Radius (m)	59	67	100	185

In one frame $r = 100$ m at a time of $t = 0.016$ second after the explosion and air density at that altitude was $\rho = 1.1 - 1.2$ kg m^{-3}. See Table 3.1 and Fig. 1.8a. Substituting these values in Eq. 3.2 gives an estimated energy release of $E = 4 \times 10^{13}$ J, which is equivalent to 1000 tons of TNT explosion of about 4.2×10^{12} J, which is an indication Trinity fission test bomb had a yield of 10 kilotons TNT based on the above calculation. The actual test bomb yield was 18–22 kilotons. Even closer values can be obtained from other frames.

See http://en.wikipedia.org/wiki/Nuclear_weapon_yield.

Still this estimate is remarkably close to the reality, given the crudeness of our analysis. Taylor (1950) [1] carefully has shown the numerical solution and noticed that the particle velocity distribution behind the shock as a function of similarity variable was quite close to linear as it is depicted in Fig. 3.2, particularly near the center of the blast. His assumption was toward particle velocity to form a solution, which is the sum of a linear term and nonlinear correction term in the similarity variables, and then he was able to explicitly determine this term by making use of the governing equations and the Rankine–Hugoniot conditions. This enabled him to find an approximate closed form solution of the entire problem, which was in error in comparison with the numerical solution by less than 5 %.

As it is stated above, Taylor (1950b) [3], in his second publication, was able to check the power law of $R \approx t^{2/5}$, to show comparison with the shock trajectory that was obtained, experimentally, from the Trinity fission bomb explosion in New Mexico. The agreement of the two various values for adiabatic index $\gamma = C_p/C_v$ was remarkably good.

In this comparison, photographs were used to measure the velocity of the rise of the slowing center of the heated volume. This velocity was found to be 35 m s^{-1}. The hemispherical explosive ball behaves like a large bubble in water until the hot air suffers turbulent mixing with the surrounding cold air. The vertical velocity of this "equivalent" bubble was computed from this analysis and was found to be 35 m s^{-2}.

"While Taylor (1950) [1, 3] was quite aware of the advantages of a Lagrangian approach to the problem, he was rather skeptical of its practicality since, as he remarked, that would introduce great complexity, and, in general, solutions can only be derived by using step by step numerical integration" of the full system of nonlinear PDEs. Actually, as a particle crosses the shock, it has an adiabatic relationship between pressure and density corresponding with the entropy, which is, endowed upon it by the shock wave during its passage past it. This naturally suggests a Lagrangian approach wherein the Lagrangian coordinate is, defined as one which retains its value along the particle path. Indeed, this matter was raised much later again by Hayes (1968) [5] who tried to contradict the suggestion by Zel'dovich and Raizer (1967) [4] that the Lagrangian formulation is as convenient as the Eulerian, even more so for the problems of blast wave type. He argued that the basic differential equation to be, solved numerically is in a non-analytic form in the Lagrangian formulation and would therefore pose difficulties, a view in agreement with Taylor's apprehension" [2].

Further analysis of this matter could be found in the classical text by Sachdev [2]. In the same reference equations involving shock wave, exact solutions of spherically symmetric flows in Eulerian coordinates are presented with great details. The exact solution is for one-dimensional gas dynamic equation, which shades light on the structure of the solutions, the blast wave being one class of solutions of these equations, where self-similarity approach is taking place. He also has suggested even exact solutions of gas dynamic equations in Lagrangian coordinates, where the approach is quite distinct and applies to all geometries—planar, cylindrical, and spherical. The basic idea behind this approach is to use the singlesecond-order nonlinear partial differential equation governing the Eulerian coordinate with the Lagrangian coordinate enthalpy h and time t as independent variables.

These solutions depend upon an arbitrary function, which is related to the entropy distribution in the gas. Applications of isentropic and non-isentropic solutions include flows with shocks of finite and infinite strength and vacuum fronts. This presentation is as follows:

$$h = \int_{y(0,t)}^{y(h,t)} r^{n-1}\rho(r,t)dr \quad n = 1, 2, 3, \tag{3.3}$$

where $y(h, t)$ is the radius of the particle with Lagrangian coordinate h at time and $n = 1, 2, 3$ for planar, cylindrical, and spherical symmetry, respectively. In the latter two cases, y represents the distance from the axis and center of symmetry, respectively. However, for complete details of this approach, refer to Sachdev book [2].

As a final note for Taylor's explosion problem, as a warning and drawback, you should remember mathematical functions only take dimensionless arguments. This is shown by power series expansions:

$$\begin{aligned} f(\xi) &= e^{\xi} \\ &= 1 + \xi + \frac{1}{2}\xi^2 + \cdots \end{aligned} \tag{3.4}$$

In this case, the leading term is obviously dimensionless, and all terms added to it must be also. In general, a function has terms of many different orders, which must be dimensionless to add up:

- Some ratios of variables and their derivatives can lead to ambiguous cases, like the ideal pendulum that is presented at the Appendix A of this book as

$$\omega_{\mathrm{B}} = \sqrt{\frac{g}{\theta}\frac{d\theta}{dz}}. \tag{3.5}$$

  This equation is dimensionally correct for *any* substitution for θ.
- Derivatives and ratios are indistinguishable to a dimensional analysis, since g/z has the same dimensions as dg/dz.
- Dimensional analysis is an aid to insight, thus it cannot completely describe the physics.

In conclusion, the creation and performance of shock waves have ken the focus of study by many engineers and scientists working in topics related to continuum physics. Shock waves—either in their weak form (acoustic waves) or their moderate to stronger form—play an important part in scientific and engineering calculations whether their existence is desirable or not.

3.3 Self-Similarity and Guderley Problem

The study of converging spherical and cylindrical shock waves in a homogeneous and symmetrical mode is of importance due to its applications in the field of nuclear engineering such as controlled thermonuclear fusion, cavitations, and blast waves. Although Guderley (1949) [8] was the first author among others demonstrating such study in a perfect gas situation, similar technique was used in nuclear fission bomb fabrications both during the Manhattan Project and later on in the design of superbomb for thermonuclear fusion process.

The creation and performance of shock waves have ken the focus of study by many engineers and scientists working in topics related to continuum physics. Shock waves, either in their weak form (acoustic waves) or their moderate to stronger form, play an important part in scientific and engineering calculations whether their existence is desirable or not. For example, in the case of gas pipelines, a sudden valve closure or opening (or any other blockage or leak) creates a response signal in the form of shock or expansion waves, whose speeds depend on the aerothermodynamic state of the gas. The change in properties behind such waves should be taken into consideration for designing the pipeline as well as the surrounding installations (for safety considerations).

For internal combustion engines, the sudden opening and closing of valves create a continuous stream of shock or expansion waves, interacting and moving down the muffler as well as other ducts. This stream has to be controlled and optimized for environmental protection (Matsumora 1993) [9].

In the field of interior ballistics of guns, the existence of shock waves ahead of and behind the projectile is an unavoidable side effect to contend with, and designs are made to divert the blast and reduce its noise level (Phan 1991) [10]. Another important military application of cylindrical converging shock waves are the generating of partly converging and partly advancing shock waves in shaped charges for armor piercing. In this case, the important usage of cylindrical converging shock waves lies in production of localized high gas pressure and enthalpies. Theoretically, area convergence is expected to strengthen the shocks, thus producing infinitely dense amounts of energy at the center of convergence (point of collapse).

The common factor here is the need for accumulating great amounts of energy in virtually point-size domains. This fact probably explains why the technical steps required to create converging shocks have their inherent difficulties.

The converging shock waves via a powerful spherical and cylindrical compression in the neighborhood of the center of the sphere and of the cylinder axis that originally was studied by Guderley are the first examples of a class of self-similar solutions of second kind. However, Stanyukovich (1969) [11] first developed an approximate method for obtaining the similarity exponent analytically.

For the case of cylindrical converging shock waves, one of the main issues faced in establishing the physical process is the shock stability. This is defined as the ability of the generated shocks to retain their required symmetric shapes if subjected to perturbations due to geometrical or physical irregularities, which is inevitable in practical considerations. Unlike plane shocks, which retain their shape due to transverse waves [12–15], two contradictory processes affect cylindrical waves: stabilizing effect due to the transverse waves and the increase in shock speed associated with the reduction in the frontal area. Therefore, the measure of stability for converging shocks should aim at minimizing the ratio between the magnitude of unavoidable perturbations and the mean value of the shock radius.

The need to stabilize the shock for as long as possible requires an efficient method for simulating the shock performance throughout the implosion process. In this respect, research activities have been diversified according to the available theoretical and technical facilities.

The "Classical Guderley Problem" [8] is considering an infinitely strong, symmetric, and homogeneous shock wave focusing on either center (or point) of spherical geometry or axial of cylindrical geometry shape. Although he did not discuss the source of generating the shock, for solving this classical problem, however, the initial state of the gas into which the shock wave is propagating is well defined and described and denoted by sets in Eq. 3.9 below, in one-dimensional Eulerian space of r-coordinate (i.e., *spherical* and *cylindrical* geometry). His assumption for the perfect gas was under perfect inviscid gas conditions. The inviscid flow is a schematic representation of the motion of mobile media such as gaseous or liquid and as well as solids under the rapid action of high pressures, which is the main theoretical model for many fields of modern technology.

Guderley demonstrated that strong cylindrical converging shock waves propagate according to a power-law relation as described below, when approaching the center and that their Mach numbers reach infinite values at the point of collapse. However, this is not possible in reality due to the effect of viscosity and heat conduction. Lighthill [16], Butler [17, 18], Stanyukoouvich [11], and Whitman [18] conducted subsequent studies under the same assumptions.

The sets of Eulerian conservation equations for inviscid flow or gas conditions are as follows, and they are valid if the viscosity and thermal conductivity of the fluid or gas are ignored:

$$\rho \frac{d\vec{u}}{dt} = \rho \vec{F} - \text{grad } p \qquad \text{Conservation of momentum,} \tag{3.6}$$

$$\frac{1}{\rho}\frac{d\rho}{dt} = -\text{div } \vec{u} \qquad \text{Conservation of mass,} \tag{3.7}$$

$$\rho \frac{d}{dt}\left(e + \frac{u^2}{2}\right) = \rho \vec{F}\, \vec{u} - \text{div } p\vec{u} + \rho q \qquad \text{Conservation of energy.} \tag{3.8}$$

In all these three sets of equation, variables and parameters expressed in them are defined as follows:

$\vec{u}$ = Velocity of gas in vector form
p = Pressure quantity
ρ = Density of fluid or gas
e = Specific internal energy

The above four elements are measured at the point within a fluid or gas, where they are continuous in that space.

Equation 3.6, also known as *Eider's equation*, relates fluid particle acceleration within an element of volume of that fluid to an external body force $\vec{F}$ and the pressure force applied on the side of the neighboring fluid particles. This equation is the generalized form of Newton's second law, which is in classical mechanics; we know it as the conservation of momentum as applied to the motion of fluid particles.

Equation 3.7 is basically expressing the law of mass conservation, which is indicating that the rate of change of density of a fluid particle is equal, with the sign reversed, to the rate of change of volume.

Equation 3.8 is expression for the law of energy conservation, which is describing change in the internal energy e and kinetic energy $\frac{1}{2}v^2$ of a fluid or gas particle as a result of an action of impressed mass forces $\vec{F}$ and surface forces (i.e., pressure p) and to an inflow of heat with intensity q from an external source.

Denoting physical flow variables in the unshocked region by the subscribed zero that is depicted in Fig. 3.3, the initial state is then expressed as

$$\begin{cases} u_0(r,t) = 0 \\ \rho_0(r,t) = \text{constant} \\ p_0(r,t) = 0, \end{cases} \tag{3.9}$$

where r denotes position ($r \geq 0$) and t time in interval of ($-\infty < t < 0$) for the converging shock wave mode and for the interval of ($0 < t < +\infty$) for the reflected shock wave mode, while u velocity, ρ mass density, and p material pressure.

Note that in Fig. 3.3, $R_s^-(t)$ is designation for converging shock wave trajectory, while $R_s^+(t)$ is trajectory for reflecting shock wave, and space-time regions are 0, 2a, 2b, and 3.

Fig. 3.3 Notional representation of converging and reflecting shock trajectory

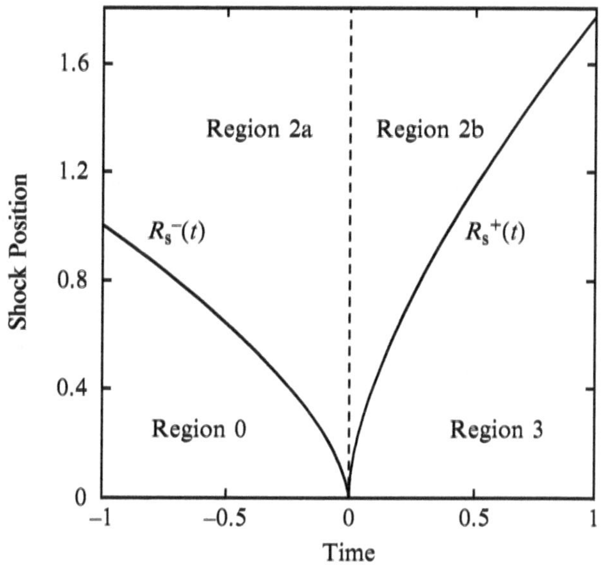

The basic sets of conservation equations (Eqs. 3.5 through 3.8) of mass, momentum, and energy govern adiabatic flow for Guderley problem, where we have smooth flow free of viscosity, heat conduction, radiation, and body forces; the one-dimensional Eulerian equations that are describing fluid motion as all continuous (i.e., non-shock) are expressed as

$$\frac{\partial \rho}{\partial t} + \frac{\partial(\rho u)}{\partial r} + (m-1)\frac{\rho u}{r} = 0, \tag{3.10}$$

$$\frac{\partial u}{\partial t} + u\frac{\partial u}{\partial r} + \frac{1}{\gamma}\left[\frac{a^2}{\rho}\frac{\partial \rho}{\partial r} + 2a\frac{\partial a}{\partial r}\right] = 0, \tag{3.11}$$

$$\frac{\partial a}{\partial t} + u\frac{\partial a}{\partial r} + (\gamma-1)\left[\frac{\partial u}{\partial r} + \frac{(m-1)u}{r}\right] = 0, \tag{3.12}$$

where a is expressing the local speed of sound and defined through the pressure and density by

$$a^2 \equiv \gamma\frac{p}{\rho}. \tag{3.13}$$

Here, we are considering only a polytropic gas with the incomplete equation of state, which is given by the following equation known as Mie–Gruneisen type as

$$p(\rho, e) \equiv \rho e \Gamma\left(\frac{\rho}{\rho_0}\right) = (\gamma - 1)\rho e. \tag{3.14}$$

Again, the symbol of e is the specific internal energy, and γ denotes the adiabatic index of fluid or gas in range of $1 < \gamma < \infty$ and m the space dimension $m = 1, 2, 3$ for one-dimensional planar, cylindrical, or spherical geometries of symmetry and homogeneous of the shock wave. The symbol of $\Gamma(\rho/\rho_0)$ is the Gruneisen coefficient and for a perfect gas is a constant and is equal to $\gamma - 1$ [18].

Note that Eqs. 3.10 through 3.12 are not valid, globally, though shock jump conditions are available to connect the pre-shock and post-shock flow field. In particular, since the converging shock wave is assumed to be an infinitely strong limit to Rankine–Hugoniot jump conditions, they may be used to connect the flow just behind and ahead of the shock front, and they can be written as

$$\begin{cases} \dfrac{\rho_{2a}}{\rho_0} = \dfrac{\gamma + 1}{\gamma - 1} \\ u_{2a} = \dfrac{2}{\gamma - 1}\dot{R}_s^-(t) \\ p_{2a} = \dfrac{2}{\gamma - 1}\rho_0\left[\dot{R}_s^-(t)\right]^2 \end{cases}. \tag{3.15}$$

Equation 3.15 sets are valid for all value of $t < 0$ where the convergent mode and subscript $2a$ as per Fig. 3.3 denote the state just behind the converging shock along with the symbol of $\dot{R}_s^-(t)$ in presentation of converging shock [19].

After shock focus and subsequent reflection about the axis or point of symmetry as well as analogous to reflection from a rigid wall in one-dimensional planar symmetry, these equation cease to be valid. Ramsey et al. [19] show the detail analysis of their novel approach to the Guderley [8] solution and argue the case for $t > 0$.

Theoretical handling of the one-dimensional form of governing equations continues to this day, with the introduction of new equations of state or constitutive relations to simulate shock dynamics in more complicated physical situations or in other types of continua (including real gases).

Some other techniques were developed by Chester–Chisnell–Whitham, which is known as the C–C–W [20–22] theory and along with Whitham's easy-shock theory [23, 24], providing researchers with a new graphically operable tool to simulate shock dynamics.

For the C–C–W theory, shocks are considered as discontinuities between continuously varying sections of fluids. The continuous fluid sections were solved for by using the modified set of quasi-1-D Euler equations in its characteristics form, while the variations across wave fronts are governed by the Rankine–Hugoniot relations.

The ultimate result was a new governing equation relating the local duct area (at shock location) to the local shock Mach number, incorporating the newly, defined Chester function, named after its inventor. Although the quasi-1-D derivation was meant to deal with flows in ducts with varying cross sections, the simplified set of equations was used for handling the cylindrical and spherical converging shocks (where symmetry ensures one-dimensionality) [25, 26].

This method of solution was extended to multiple dimensions using the ray–shock theory deduced by Whitham [23, 24]. Based on concepts from geometrical acoustics, the method employs successive shock contours and their orthogonal trajectories (rays) as curvilinear coordinate lines. It was assumed that no lateral flow takes place across the ray lines, meaning that the rays coincide with streamlines at the shock location. The geometrical compatibility requirements lead to one differential equation relating the shock Mach number M and the ray tube area, A, for each tube. A second relationship between the two quantities is deduced using the C–C–W theory for the motion of a shock wave down a tube of varying cross section. The resulting equations are of hyperbolic nature, and a solution can be deduced using the method of characteristics, which also describes the motion of lateral waves on the shock front. These are interpreted as the intersection of acoustic waves with the shock front, and the case where these waves break is termed as "shock-shock" (which is well visualized in the case of Mach reflection) [27].

More details can be found in reference by El-Mallah [27] in his thesis. In summary, the problem of collapse of a spherical or cylindrical cavity and converging shock wave from a spherical or cylindrical as a result of implosion are considered to be self-similar solutions, and it can be shown to be unstable for most ranges of γ.

A related problem is that of converging shock waves, which also possesses a similarity solution. Indeed, self-similar solutions of both these problems belong to the class called the "second kind" (Zel'dovich and Raizer (1967)) [4] for which dimensional analysis or group properties of the PDEs do not fully determine the self-similar form of the problem; they require a global solution of an eigenvalue problem for the reduced system of ODEs. Typically, for this class of problems, the exponent in the definition of the similarity variable turns out, in general, to be an irrational number.

For the converging shock problem, which was first studied by Guderley (1942) [8], this exponent in the similarity variable $\xi = rt^{-\alpha}$ was found to be 0.717 for the spherical converging shock for $\gamma = 1.4$. Several other investigators later refined this value, and good discussion is given by Sachdev [28].

3.4 Physics of Nuclear Device Explosion

There are two atomic bombs that were designed and went to production as a result of the Manhattan Project, the first one called with the code name of *Little Boy* and the second one called with the code name of *Fat Man*. See Fig. 3.4a, b.

Between the two bombs, the Little Boy had a simpler design than Fat Man model with a code name Gadget that was tested at Trinity site in New Mexico, and famous

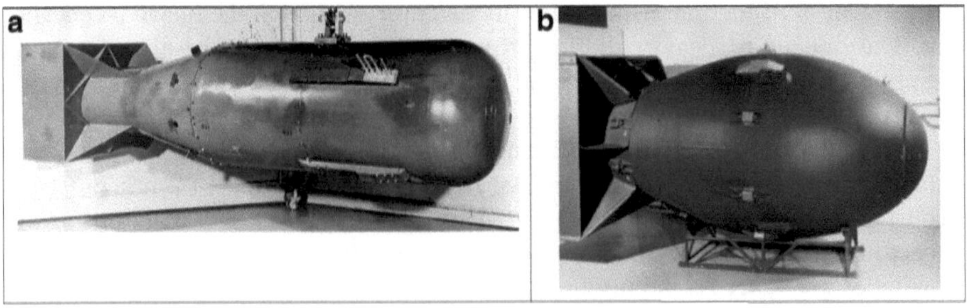

Fig. 3.4 Photo image both from World Word II bomb: (**a**) Little Boy image; (**b**) Fat Man image

Taylor (1950a) [1] calculation for the formation of a blast wave by very intense explosion was published. This analysis was derived from the aftermath of that explosion, which the solution does fall into the category of *self-similarity of first kind* as it was discussed previously. At this stage we are going to explain the design mechanism and physics of both bomb and how they did work and how the self-similar solution of the problem of the implosion and explosion is applied, in particular case for the Fat Man plutonium bomb, and how the converging and diverging shocks turns the problem into self-similar of first and second kind.

3.4.1 Little Boy Uranium Bomb

The Little Boy uranium bomb triggered a nuclear explosion, rather than implosion, by firing one piece of Uranium-235 into another. When enough U^{235} is brought together, the resulting fission chain reaction can produce a nuclear explosion. However, the critical mass must be assembled very rapidly; otherwise, the heat released at the start of the reaction will blow the fuel apart before most of it is consumed.

To prevent this inefficient pre-detonation, the uranium bomb uses a gun to fire one piece of U^{235} down the barrel into another. The bomb's gun barrel shape was believed to be unquestionably reliable and had never been tested. In fact, testing was out of the question since producing Little Boy had used all of the purified U^{235} produced to date; therefore, no other bomb like it has ever been built.

Detonated by a mechanism that resembled a cannon, Little Boy had a muzzle or target that was a hollowed-out subcritical mass of uranium. The cannon ball was another subcritical mass of uranium, which fit perfectly into the hollow of the target as a plug. The plug was propelled down the cannon barrel by several thousand pounds of high explosive. When it hit, the combination of compression and increased mass pushed the uranium to the supercritical level and the bomb went off.

Figure 3.5a–c is a possible schematic and geometrical configuration of Little Boy, with a total assembled weight of 9700 lb, length of 10 ft, and diameter of 28 in.

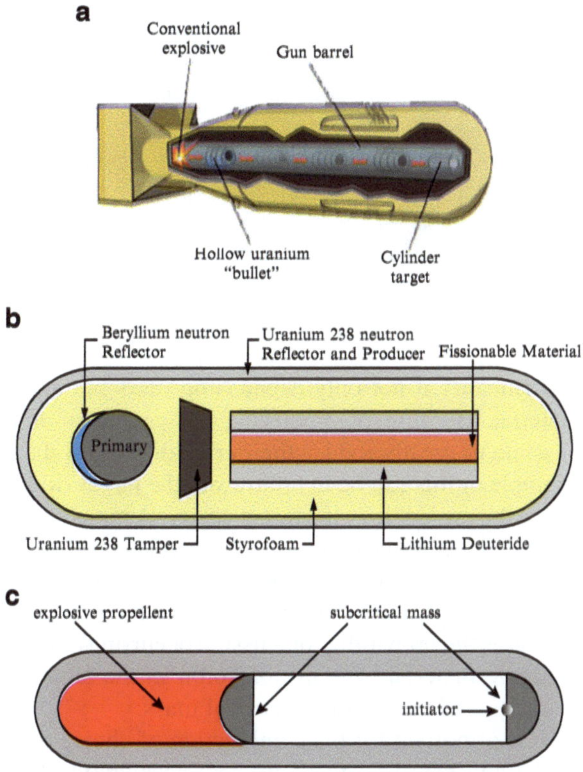

Fig. 3.5 Overall interior design of Little Boy bomb

In this bomb the source of nuclear explosion is highly enriched Uranium-235 equivalent to 15,000 tons of TNT.

As stated above, the Little Boy was a gun-type bomb. The gun assembly design is simplified as much as possible, in order to reach mass criticality for fission process of U^{235}. In this design for better efficiency, a tamper of U^{238} was chosen, while for other elements, beryllium was used as a neutron reflector. See Fig. 3.5b.

Figure 3.6 shows generic schematic of gun assembly of Little Boy as a type of weapon, with the fissile material that is divided into two subcritical pieces: one that is known as the projectile that is propelling to the other one which is known as the subcritical target. This process takes place as a result of the pressure of propellant combustion gases in a gun barrel. As it is demonstrated in Fig. 3.6, the "singular-gun" design contains a single projectile, which is accelerating to a singular stationary subcritical target, and of course limits the performance of the device. However, the performance of such design can be greatly improved by accelerating both subcritical pieces toward each other, a process that is called as "double-gun" design.

If the objective is to assemble a supercritical mass of fissile material, then the theoretical higher limit of the quantity of fissile material in a gun design is a little bit under two critical masses. This is because it is necessary for the projectile and target to remain subcritical before the device operates.

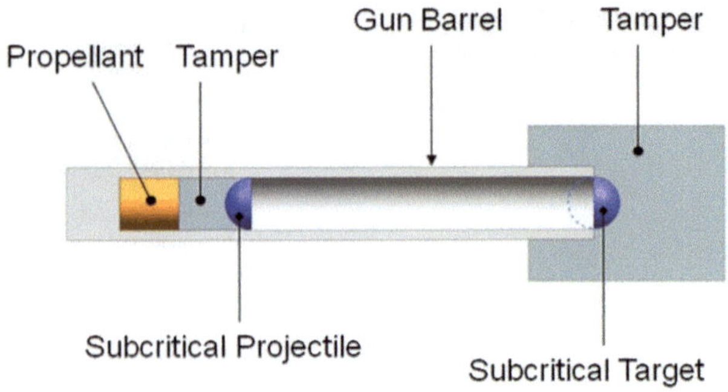

Fig. 3.6 Generic schematic of singular-gun design

Criticality it not only depends on mass but also on the geometry of the fissile material. Because of modifying the geometry of the target and projectile, it is possible to greatly reduce their criticality (until the point of system assembly). An example of this can be to hollow out the target (to greatly reduce its criticality) and to fire the projectile to the target cavity. A more optimum geometry, thus, assembles more than the initial configuration could have allowed. This methodology can allow the use of three times the bare sphere critical mass of fissile material in a gun design.

The use of a neutron moderator (a medium which slows down neutrons and increases the probability of a fission occurring) can also increase the allowable mass of fissile material. The use of boron in the target and/or projectile can moderate neutrons and thus raise the critical mass.

It is important for the combination of choice of propellant and length/thickness of the gun barrel to give sufficient acceleration to the projectile before insertion. Because it is desirable to minimize weight and length of the weapon, insertion velocities are limited to velocities, which are below 1 km s^{-1} [29].

Because it is probable that the target and projectile will be close to a critical mass before assembly, it is probable that a critical configuration will be obtained before the complete insertion of the projectile in the target or possibly before the projectile reaches the target. The probability of this occurring increases at the same time as the mass of the target and projectile increases. As soon as a critical mass assembles, there is a chance of pre-detonation. Thus, it is desirable to have as high insertion velocity as possible to minimize the risk of a "fizzle" [29].

Because of the inherent pre-detonation risk during assembly, the choice of fissile material is limited. Plutonium undergoes spontaneous fission and thus is a neutron source itself—this greatly raises the risk of pre-detonation. This is not the case for uranium, and so it becomes the preferred material. However, utilizing natural or depleted uranium for the tamper around the target can provide significant background neutrons. Thus it is necessary to avoid them. Implosion devices do not suffer in the same manner, because the implosion timescale is a lot shorter than the assembly time for a gun device [29].

Figure 3.7 shows overall design aspect of Little Boy and its cross section with two subcritical pieces at each end of the gun barrel that are shown as U^{235} projectile rings and U^{235} target rings, respectively.

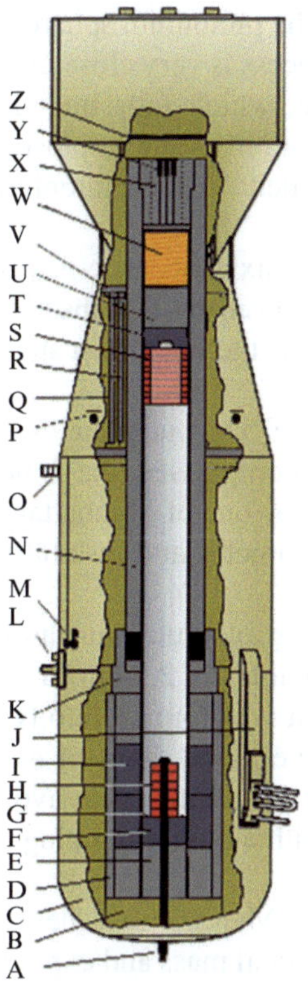

Cross-section drawing of Y-1852 Little Boy showing major mechanical component placement. Drawing is shown to scale. Numbers in () indicate quantity of identical components. Not shown are the APS-13 radar units, clock box with pullout wires, baro switches and tubing, batteries, and electrical wiring. (John Coster-Mullen)

Z) Armor Plate
Y) Mark XV electric gun primers (3)
X) Gun breech with removable inner plug
W) Cordite powder bags (4)
V) Gun tube reinforcing sleeve
U) Projectile steel back
T) Projectile Tungsten-Carbide disk
S) U-235 projectile rings (9)
R) Alignment rod (3)
Q) Armored tube containing primer wiring (3)
P) Baro ports (8)
O) Electrical plugs (3)
N) 6.5" bore gun tube
M) Safing/arming plugs (3)
L) Lift lug
K) Target case gun tube adapter
J) Yagi antenna assembly (4)
I) Four-section 13" diameter Tungsten-Carbide tamper cylinder sleeve
H) U-235 target rings (6)
G) Polonium-Beryllium initiators (4)
F) Tungsten-Carbide tamper plug
E) Impact absorbing anvil
D) K-46 steel target liner sleeve
C) Target case forging
B) 15" diameter steel nose plug forging
A) Front nose locknut attached to 1" diameter main steel rod holding target components

"Atom Bombs: The Top Secret Inside Story of Little Boy and Fat Man, "2003, p 112,
John Coster-Mullen drawing used with permission

Fig. 3.7 Cross-sectional drawing of Y-1852 Little Boy

Neutronic analysis for mass criticality for efficiency purpose is presented in the following subsection. In this section, we have briefly discussed the physics and mathematics that drives such criticality. However, for readers who are interested in more details of such analyses, Reed (2015) [30] has published a classical book in public domain that reader can refer to it.

3.4.2 Fat Man Plutonium Bomb

The Fat Man plutonium bomb had a more complex design of the two and designed with geometrical configuration of bulbous shape with length of 10 f. 8 in. and diameter of 60 in. with total assembled weight of 10,800 lb. The physics of this design was based on bomb containing a sphere of the metal Plutonium-239, and it was surrounded by blocks of high explosives that were designed to produce a highly accurate and symmetrical implosion. This would compress the plutonium sphere to a critical density and set off a nuclear chain reaction. The process is very close to an ideal situation of implosion process in symmetrical mode, which falls into the category of *self-similarity of second kind* that drives the solution of the implosion problem. Scientists at Los Alamos were not entirely confident in the plutonium bomb design, so they scheduled the Trinity test.

The basic structure of the Fat Man was based on a series of six concentric-nested spheres. The outermost was the explosive lens system, followed by the absorber shell, the uranium reflector shell, the plutonium pit, and lastly, the innermost shell, the neutron initiator.

The shell system, basically, worked as an implosion device. The outer shell was made of high-powered explosive that, when detonated, compressed the inner spheres and charged the uranium. The design was suggested by one of Manhattan's project scientists, Robert Christy, in order to minimize asymmetry and instability problems during implosion.

The innermost shell, the plutonium pit, contained 6.2 kg of plutonium alloy contained in a 9.0-cm shell. It was solid except for an approximately 2.5-cm cavity in the center where the neutron initiator was placed. The pit was formed into two hemispheres. Since plutonium is a chemically very reactive metal, as well as a significant health hazard, each half sphere was electroplated with nickel to avoid deteriorating reactions. The pit had a 2.5-cm hole capped with a plutonium plug to allow the insertion of the initiator after assembly.

Figure 3.6 is an overall schematic of the interior of Fat Man bomb, where it clearly shows an implosion process for the bomb to go on critical mass and explode at its nuclear fission energy (Figs. 3.8 and 3.9).

3.4.3 Problem of Implosion and Explosion

In order to consider this problem and the possible integration of it into the physics of nuclear device explosion mechanism, we need to have some fundamental understanding of self-similar motion of spherical symmetry in particular. Using self-similar method for motions of spherical, cylindrical, and plane waves in a gas, it was understood by many scientists and researchers in the past, and we look at one-dimensional motions of a fluid determined as motions whose characteristics depend only on a single geometrical coordinate (i.e., r in case spherical and cylindrical shape) and on time t.

As it has been stated in the previous two chapters of this book, Sedov [7], Guderley [31], Taylor [1], and others have tackled this problem independent of each other within various closed time. Basically what they have shown is that in one-dimensional motions, which are produced by spherical, cylindrical, and planar waves, the method of dimensional analysis and similarity theory leads the problem of nonlinear to an exact solution for problems of unsteady motion of a compressible fluid. This type of approaches by finding the exact solutions obtained might be

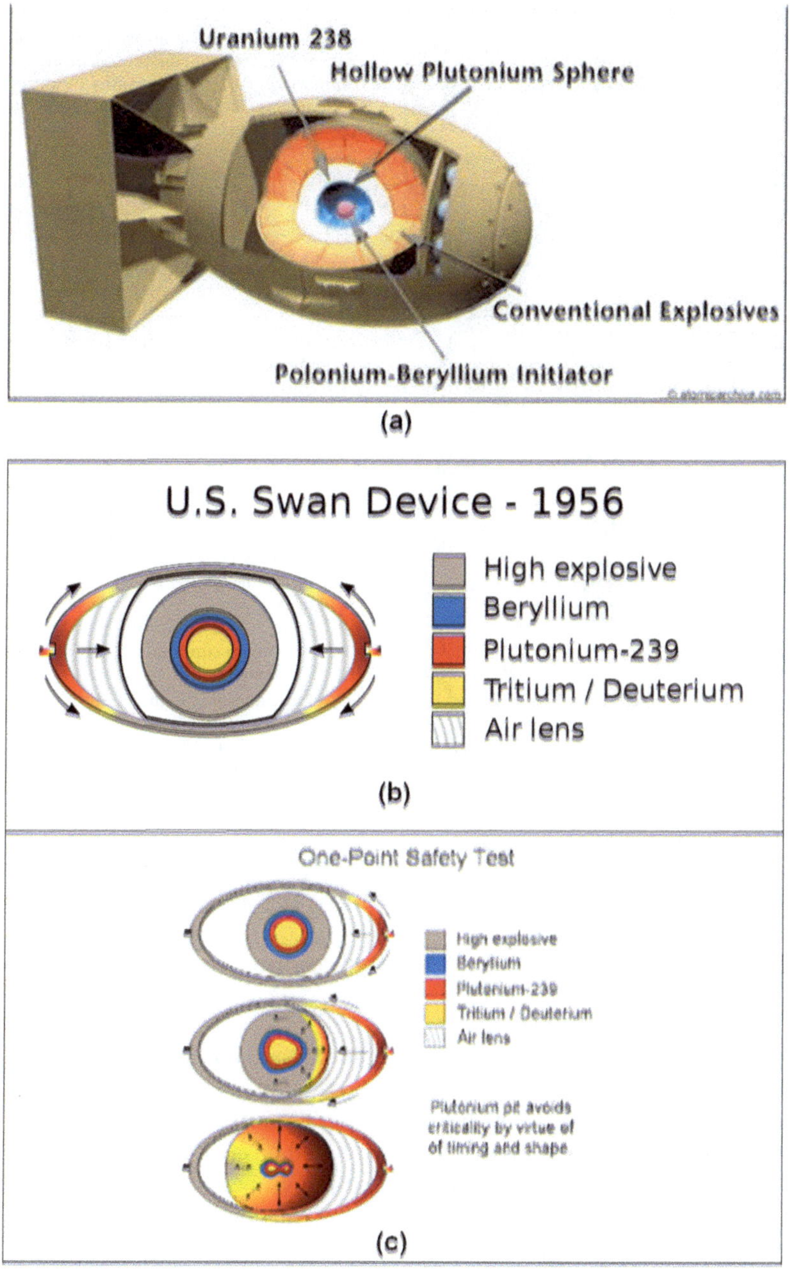

Fig. 3.8 Implosion process of Fat Man bomb

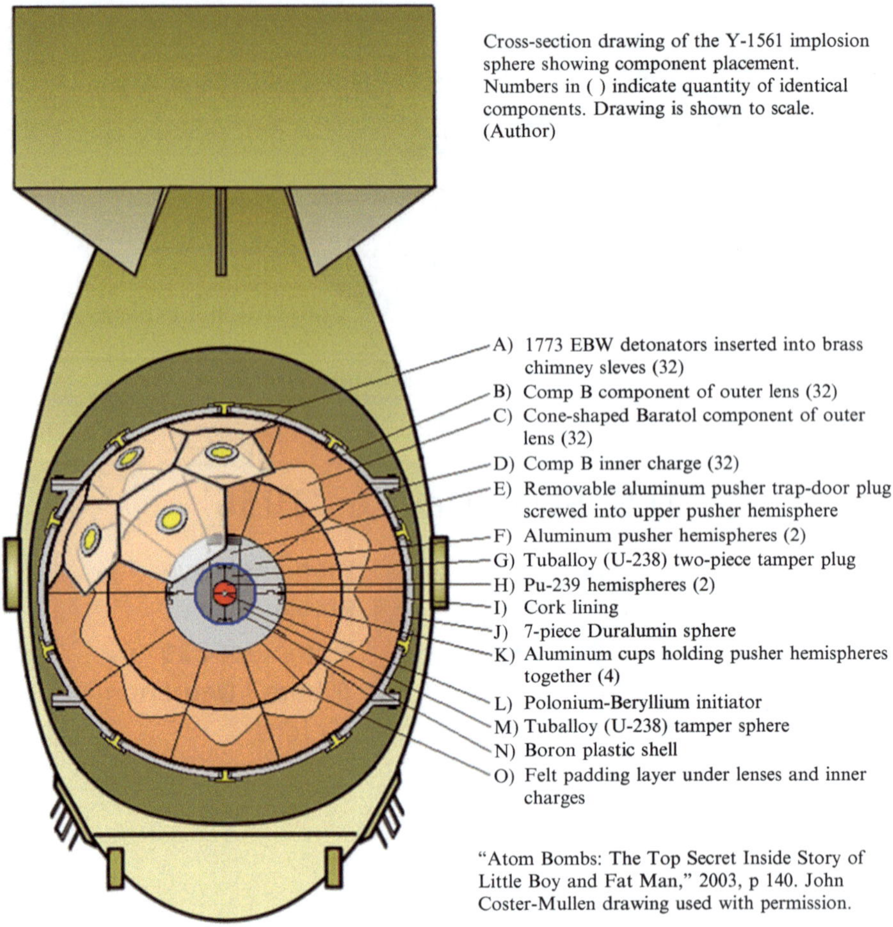

Fig. 3.9 Cross-sectional drawing of Y-1561 Fat Man

helpful to confirm the accuracy of various approximated solutions of the problem in fluid dynamics.

For this matter, we consider the characterization of the problems that can be solved by dimensional analysis and similarity methods; we can consider a suitable function and characteristic parameter describing the one-dimensional motion within the Eulerian system. From this condition viewpoint, the main suitable/desired function has variables of velocity υ, density ρ, and pressure p, and the characteristic parameters as we stated are the linear coordinate r and time t. This assumption is involving another characteristic among the ones above as a constant a, with the dimension that at least contains the symbol of mass M. Therefore, putting constant characteristic of a can its dimension perspective without loss of generality, we can write it as

$$[a] = ML^k T^s. \tag{3.16}$$

Thus, for the unknown functions of velocity, density, and pressure, we can establish the following relationships as

$$v = \frac{r}{t}V \qquad \rho = \frac{a}{r^{k+3}t^s}R \qquad p = \frac{a}{r^{k+1}t^{s+2}}P, \tag{3.17}$$

where V, R, and P are abstract quantities, and, therefore, they depend only on nondimensional combinations including r, t, and other parameters involved in the problem of interest in hand. Generally speaking, these characteristics are functions of two-dimensional variables, however, if among the characteristic parameters, in addition to a, there is one more individual constant b with dimension independent of a. In general, there can be many characteristic constants, but their dimension has to be dependent on a and b with possible independent dimensions with fixed exponent k, s, m, and n that can be integral, fractional, or transcendental numbers [7].

However, the actual determination of these exponents in a particular problem of interest is connected with the setup formulation of the problem and properties of unknown solutions, which always exceed the limits of dimensional theory. See Chap. 1 of this book.

Given the preceding text, since the dimension constant characteristic a is depending on symbol of mass M, then without again, loss of generality, we can always present the constant b so that its dimension will not contain the element of mass symbol M as

$$[b] = L^m T^n. \tag{3.18}$$

In this case, $r^m t^n / b$ will lead to only nondimensional combination, which for $m \neq 0$ can be replaced by the variable λ as below:

$$\lambda = \frac{r}{b^{1/m} t^{\delta}} \qquad \text{where} \qquad \delta = -\frac{n}{m}. \tag{3.19}$$

However, if $m = 0$, then V, R, and P will be dependent only on time t, where in that case velocity v is proportional to r. The corresponding particular motions are studied by Sedov [7] in details; he also shows in addition to the variable parameter λ that the solution can also depend on a number of constant abstract parameters. He assumes that it is among characteristic parameters of the problem, in addition to r and time t. There are only two constants with independent dimensions. With this content in mind, then partial differential equations, which are satisfied by the velocity, density, and pressure in the unsteady one-dimensional motion of an incompressible fluid, can be replaced by a set of ordinary differential equations for the quantities V, R, and P. Solutions of these ordinary differential equations either can be obtained in exact closed form or approximated by means of numerical integration. Such kinds of motions are called self-similar types, and we now formulate problem of explosion and implosion, which can easily be solved by the method of self-similar.

Consider the continuity equation of motion and energy in ideal gas medium in absences of heat conductivity as follows:

$$\frac{\partial v}{\partial t}+v\frac{\partial v}{\partial r}+\frac{1}{\rho}\frac{\partial p}{\partial r}=0, \tag{3.20}$$

$$\frac{\partial \rho}{\partial t}+\frac{\partial \rho v}{\partial r}+(\nu-1)\frac{\rho v}{r}=0, \tag{3.21}$$

$$\frac{\partial}{\partial t}\left(\frac{p}{\rho^{\gamma}}\right)+v\frac{\partial}{\partial r}\left(\frac{p}{\rho^{\gamma}}\right)=0. \tag{3.22}$$

These sets of equation are very similar to sets of Eqs. 3.10–3.12 but written in different form, where again, γ is adiabatic index and $\nu=1$ is for the planar motion for ideal gas, $\nu=2$ for the cylindrical, and $\nu=3$ for the spherical case. Applying the arbitrary quantities V, R, and P from Eq. 3.17, we can easily find that $k=3$ and $s=2$, and in case of general relativity theory, there are two other fundamental constants such as speed of light c and the gravitational constant f that come to play. In this case $a=f$ and arbitrary quantities V, R, and P are dependent only on a quantity $\lambda=r/ct$.

A self-similar method can be in place to solve the new sets of equation based on the functions of V, R, and P so in case of strong shock in one-dimensional spherical coordinate system moving outward, we are solving Taylor's problem and for the shock going inward we are solving Guderley's problem. Sedov [7] has shown the algebraic integral solution for self-similar motions in details by and for strong shock; we have the following form by introducing a new variable z as a function of V in the form of $z(V)$ that results from relation of $z=\gamma P/R$, where it is formulated from $\Re T=(r^2/\gamma t^2)z$. Here T is the temperature and $\Re$ is the gas constant:

$$\frac{z}{R^{\gamma-1}}=C_1\left[R(V-1)+\frac{C_2}{\gamma^{\nu-\omega}}\right]^{[\omega(\gamma-1)]/[\nu-\omega]}\frac{1}{\lambda^2}, \tag{3.23}$$

where $\omega=k+3$ and C_1 and C_2 are arbitrary constant of integration. It is obvious that variables z and V as well as function of $z(V)$ are independent of indexes k, s, and m but are well determined by the type of self-similar motion of first or second kind that falls into explosion and implosion problem, respectively. Sedov [7] shows different plots of adiabatic integral paths for different conditions of point $O(z,V)$ is z and V v plane, where an asymptotic formulation induced from Eq. 3.23 is based on ω either being negative ($\omega<0$), positive ($\omega>0$) or being equal to zero ($\omega=0$). These asymptotic sets of formulation in case of $O(z=0,V=0)$ are given as

$$z = CV^2 \quad \lambda = \frac{C_1}{V} \quad \text{and} \quad z = \frac{\gamma}{\omega} V \quad \lambda = \frac{C_1}{\sqrt{V}}. \tag{3.24}$$

Note that for *nonself-similar motions*, different curves in $z - V$-Plane correspond to the gas motion at different instants. On the other hand, for *self-similar motions*, the field of gas motion in $z - V$-Plane at different instants or for different points or particles corresponds with the same curve on the adiabatic integral curve, which is corresponding to the plot of ordinary differential equations (ODEs) for the shock conditions under self-similar motions [7].

From the formulation of self-similar motion of these ODEs, it follows that the shock coordinate r in the form of $r = \lambda bt^{\alpha}$ and variable $\lambda = r/bt^{\alpha}$ at the shock are functions of time t and characteristic dimensional constant a and b.

In particular, cases of the following situation are possible: The gas motion is self-similar but the motion of boundaries of shock waves is determined by supplementary constants. Thus, the shock coordinate r depends not only on a, b, and t but also on other dimensional constants. In these cases the formula for $\lambda = r/bt^{\delta}$ approaching to constant value λ_0 at the shock is not true; thus, in correspondence with the assumed definitions, such motions, considered as a whole, will be called nonself-similar, although self-similarity is violated only on the boundary. Moreover, a nondimensional combination cannot be formed from the three quantities on a, b, and t; therefore, for the discontinuity surface, we have the following conditions [7]:

$$\lambda = \lambda_0 = \text{constant} \quad r = \lambda_0 bt^{\alpha}. \tag{3.25}$$

Consequently, in $z - V$-Plane fixed points correspond to the shocks for self-similar motions with fixed values of variables λ, R, z, P, and V. Furthermore, for the value of shock velocity c, a formula of the following form always may be written as

$$c = \frac{dr}{dt} = \alpha \cdot \frac{r}{t}. \tag{3.26}$$

In analyses of Eq. 3.26, it is obvious for self-similar motions that α is constant. For $r > 0$ and $t > 0$, the velocity phase propagation is directed outward and away from the center when $\alpha \gg 0$. Therefore, for $\alpha > 0$, the shock waves are divergent and, thus, for $\alpha < 0$, the shock waves are directed inward, and they are convergent and the velocity of phase motion decreases. If $r > 0$, the time t increases, but $t < 0$, then we have the reverse character behavior of the motion of shock waves. Figure 3.10 is showing depiction of divergence and convergence of such shock waves behavior and characteristics along with adiabatic compression or rarefaction arises in front of the core.

On parabola equation of $z = (\alpha - V)^2$, phase velocities are equal to the speed of sound; thus above this parabola, the velocities are subsonic and below it, they are subsonic. In the general case of nonself-similar motions, the abstract quantity α is a certain function of time t.

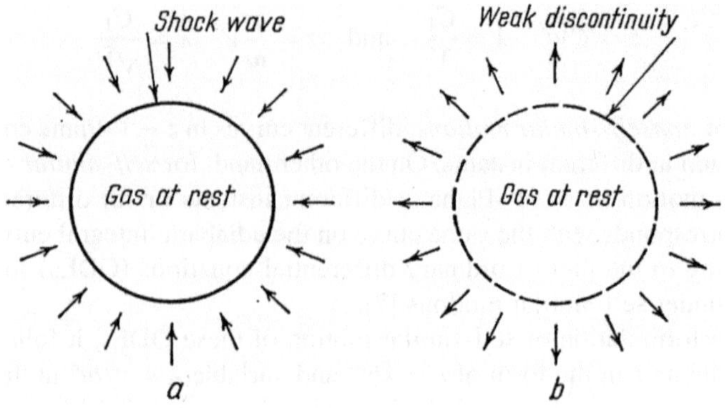

Fig. 3.10 Depiction of motion (**a**) implosion and (**b**) explosion characteristics

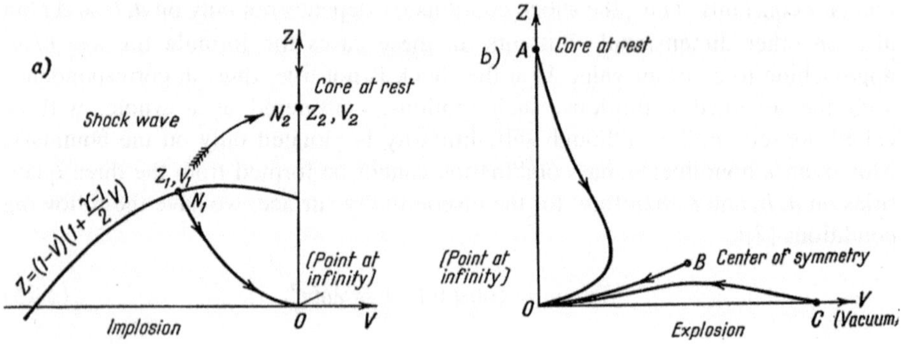

Fig. 3.11 Depiction of integral curves corresponding to (**a**) implosion at a point and (**b**) explosion from a point

For case of explosion and implosion problem at a point where the corresponding point $O(z, V)$ is placed at infinity as $O(z = 0, V = 1)$, when initial velocity, density, and pressure is uniform everywhere as it is depicted in Fig. 3.10, then $\omega = 0$ and $\alpha = 1$; the appropriate field of the integral curves in the $z - V$-plane is depicted in Fig. 3.11.

However, for points at infinity, corresponding to strong implosion or explosion, the asymptotic formulas near the point $O(z, V)$ are given as

$$z = CV^2 \quad \lambda = \frac{C_1}{V}. \tag{3.27}$$

Sedov [7], extensively, has provided the interpretations of curves in both plots of Fig. 3.11.

The first American and Soviet implosion devices were "Trinity"/"Fat Man" and RDS-1, and they were very similar in their design. The theory of operation behind these types' nuclear device and design, basically, lies within the process, when the

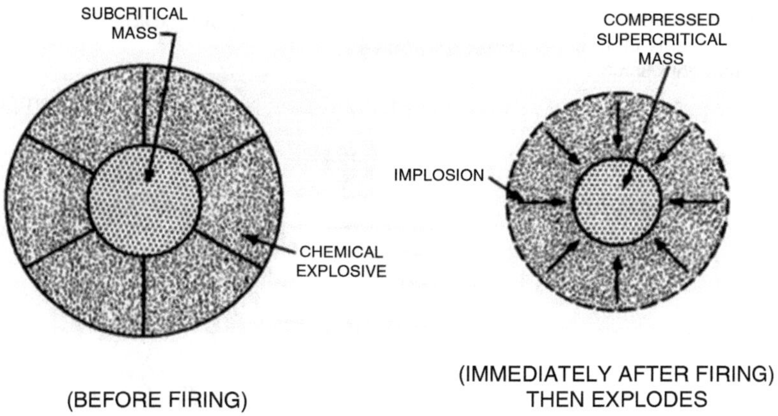

Fig. 3.12 Depiction of implosion process by detonation of TNT in nuclear device

detonated TNT compresses the plutonium into a critical mass. The critical mass then produces a nuclear chain reaction similar to the domino chain reaction (discussed in the next section). The chain reaction then promptly produces a big nuclear or thermonuclear reaction in the device. The principle physics of explosion and implosion, which was described in the previous text, in general, works with the principle concept in an implosion assembly to compress a subcritical mass of spherical or cylindrical shape of fissile material to the point where it becomes supercritical rapidly, by means of specially designed explosives. See Fig. 3.12 here.

This implosion process certainly falls in physics and mathematics of self-similar energy equation of generated shock wave motion, as it was described at the beginning of this section. This process in principle works by means of initiating the detonation of the explosives on the outer surface, in order for the instant shock waves that are generated by detonation waves of high explosive to compress the fissile material, which rapidly increases the density. Energy from these shock waves compresses the core of the device in a self-similar fashion as it is shown in Fig. 3.13, and it raises the density to the supercriticality point. This leads to a very powerful nuclear explosion. High explosives are needed to produce this large compression of the fissile material.

The artistic image of the process of self-similar method in case of plutonium core bomb (i.e., typical thermonuclear weapon or American Fat Man and Russian RDS-1) is shown in Fig. 3.13 here.

Clearly, in case of detonation of a thermonuclear initiation process, artistic form of Fig. 3.14 is a good indication of self-similar of first and second kind for shock waves' energy motion and continuity equations as well. As the figure indicates, both implosion and explosion process are taking place.

To further enhance the above process in order to achieve a rapid mass criticality and pressing the core density of fissile materials in nuclear devices, certain devices, mainly reflector mechanism, should be in place. Figure 3.15 for Fat Man design suggests such conceptual geometrical configure. The main task of these reflectors is

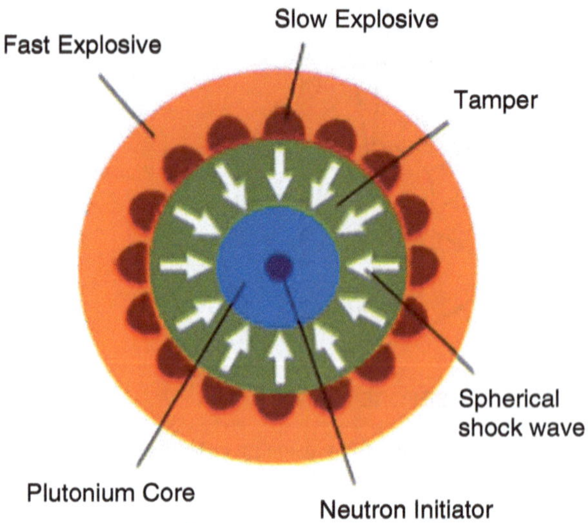

Fig. 3.13 Self-similar shock waves energy motion for plutonium core bomb

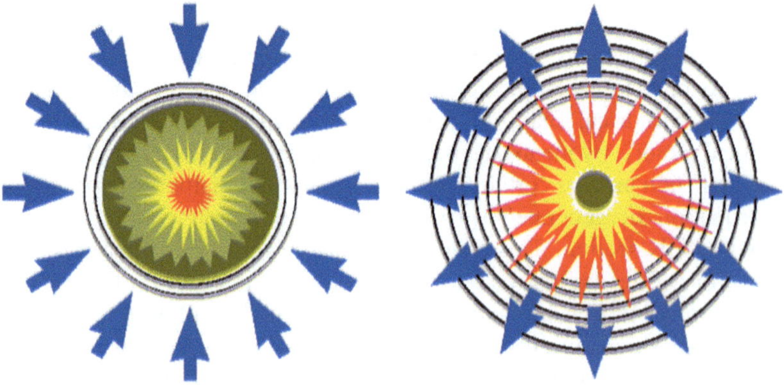

Fig. 3.14 Depiction of self-similar of implosion and explosion process

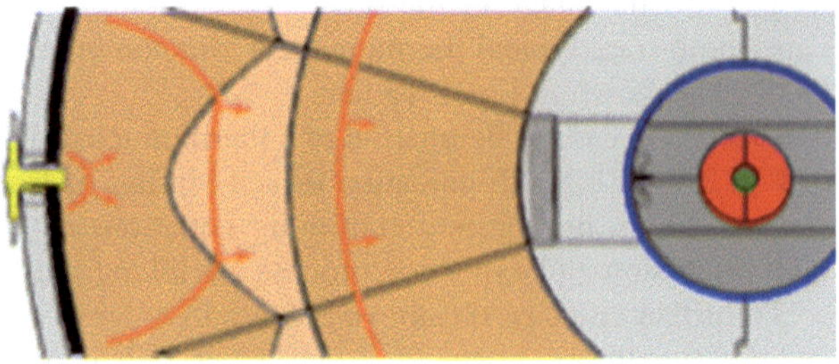

Fig. 3.15 Mechanism of reflectors

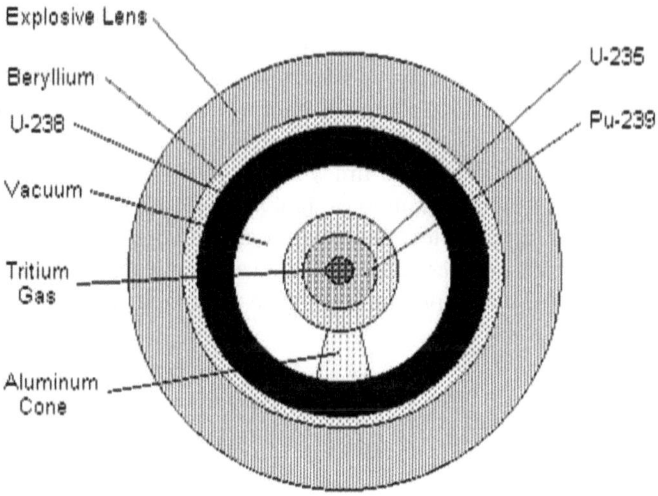

Fig. 3.16 Typical cross section of thermonuclear device

to converge the initial shock waves' energy of an implosion to compress solid uranium or plutonium with a resultant factor of 2–3 of its original mass and density via inertial process.

The compression occurs very rapidly and the window for neutron initiation lasts only a few microseconds. For maximum performance, the time of the neutron initiator must be under a microsecond. For internal neutron initiators, the initiation time is close or even simultaneous with the maximum compression. The neutron initiation theory is required; in order for nuclear weapons to function successfully, it is necessary for the fission chain reaction to initiate at the correct time. When the system becomes supercritical, a neutron is necessary in order to begin the fission process.

This "window" for successful neutron initiation differs depending on the design of the weapon. Gun assembly devices (i.e., Little Boy design) stay in a supercritical state during a relatively long time—a time which is sufficient for a background neutron to initiate the fission chain reaction. For further information, see next section here.

In case of thermonuclear weapons, Fig. 3.16 is suggesting a good indication of geometry and layout of such device to take advantage of implosion and explosion using power law of $\xi = rt^{\alpha}$ in order to solve continuity equations of second kind self-similarity (i.e., $\alpha < 0$) and continuity of first kind self-similarity (i.e., $\alpha > 0$).

In physics of laser-driven fusion via pellet of D–T, similar concept of implosion and explosion is employed. Further information is provided further down in this chapter.

Under process of laser compression of matter to superhigh densities for thermonuclear applications, Nuckolls et al. [32] is suggesting that two isotope hydrogen deuterium (D) and tritium (T) may be compressed to more than 10,000 times liquid density by an implosion system energized by a high-energy laser beam. This scheme makes possible efficient thermonuclear burn of small pellets of heavy hydrogen isotopes and makes feasible fusion power reactors using practical laser or particle beam. Such idea of compression was based on observation that was taking place at the surface of the Sun, where hydrogen in the center of the Sun is believed to exist at more than 1000 times liquid density and at pressures greater than 10^{11} atm as well as temperature of roughly 1–2 [5] KeV. In this process at the Sun's surface suggesting that electrons in white dwarf cores are Fermi-degenerate, the pressure is a minimum determined by the quantum mechanical uncertainty and exclusion principle [33]. The pressure of dense hydrogen with Fermi-degenerate electrons is [34]

$$p = \frac{2}{3} n_e \varepsilon_F \left[\frac{3}{5} + \frac{\pi^2}{4} \left(\frac{kT}{\varepsilon_F} \right)^2 - \frac{3\pi^4}{80} \left(\frac{kT}{\varepsilon_F} \right)^4 + \cdots \right], \tag{3.28}$$

where:

$n_e =$ the electron density
$\varepsilon_F = \frac{h^2}{8m} \left(\frac{3}{\pi} n_e \right)^{2/3} =$ the Fermi energy
$kT =$ the thermal energy
$h =$ Planck's constant
$m =$ the electron mass

At 10^4 times liquid density ($n_e = 5 \times 10^{26}$), the minimum hydrogen pressure occurs when $kT \ll \varepsilon_F$ and is ~10^{12} atm. Thermodynamically, the pressure applied to an implosion system is Eulerian and obeys *pdV* work-generating kinetic energy, which is converted near isentropically to internal energy concentrated in the compressed volume. Because pressure is energy per unit volume, the maximum average pressure equals the applied pressure multiplied by the compression ratio. Furthermore, additional pressure multiplication occurs at the core center of the nuclear device assembly or micro-balloon glass-containing D–T, because of convergence effects, a pure self-similar of second kind process and solutions for continuity equations [7].

3.4.4 Critical Mass and Neutron Initiator for Nuclear Devices

Scientists knew that the most common isotope, Uranium-238, was not suitable for a nuclear weapon. There is a fairly, high probability that an incident neutron would be captured to form Uranium-239 instead of causing a fission. However, Uranium-235 has a high fission probability.

Of natural uranium, only 0.7 % is Uranium-235. This meant that a large amount of uranium was needed to obtain the necessary quantities of Uranium-235. In addition, Uranium-235 cannot be separated chemically from Uranium-238, since the isotopes are chemically similar.

Alternative methods had to be developed to separate the isotopes. This was another problem for the Manhattan Project scientists to solve before a bomb could be built.

The research had also predicted that Plutonium-239 would have a high fission probability. However, Plutonium-239 is not a naturally occurring element and would have to be made. The reactors at Hanford, Washington, were built to produce plutonium.

Both uranium and plutonium were used to make bombs before they became important for making electricity and radioisotopes. The type of uranium and plutonium for bombs is different from that in a nuclear power plant. Bomb-grade uranium is highly enriched (>90 % U-235, instead of up to 5 %); bomb-grade plutonium is fairly pure Pu-239 (>90 %, instead of about 60 % in reactor grade) and is made in special reactors.

Since the 1990s, due to disarmament, a lot of military uranium has become available for electricity production. The military uranium is diluted about **25**:**1** with depleted uranium (mostly U-238) from the enrichment process before being used in power generation. For over two decades to 2013, one tenth of United States' electricity was made from Russian weapons' uranium. Military plutonium is starting to be used similarly, mixed with depleted uranium.

As previously stated, in order for nuclear weapons to function successfully, it is necessary for the fast neutron fission chain reaction to initiate at the correct time. When the system becomes supercritical, a neutron is necessary in order to begin the fission process. This "window" for successful neutron initiation differs depending on the design of the weapon. A chain reaction refers to a process in which neutrons released in fission produce an additional fission in at least one further nucleus. This nucleus in turn produces neutrons, and the process repeats. The process may be controlled (nuclear reactor power plant) or uncontrolled (nuclear or thermonuclear weapons). Figure 3.17 shows an artistic sketch of Uranium-235 chain reaction driven by neutron-splitting atom of ^{235}U as a result of release of energy, which obeys Einstein's famous theory of kinetic energy, where $E = mc^2$.

The chemical fission reaction is written as

$$^{235}\text{U} + \text{n} \rightarrow 2 \text{ or } 3\ \text{n} + 200\ \text{MeV}.$$

If each neutron releases two more neutrons, then the number of fissions is doubling in each generation. Thus, in 10 generation of neutron, there are 1024 fission and 80 generations of neutron about 6×10^{23} fissions (a mole). As a rule, in the nth generation of neutron, there would be $(2)^n$ neutrons available. This process of chain reaction then is telling us since the weight of 1 nucleus of 25 is 3.88×10^{-22}

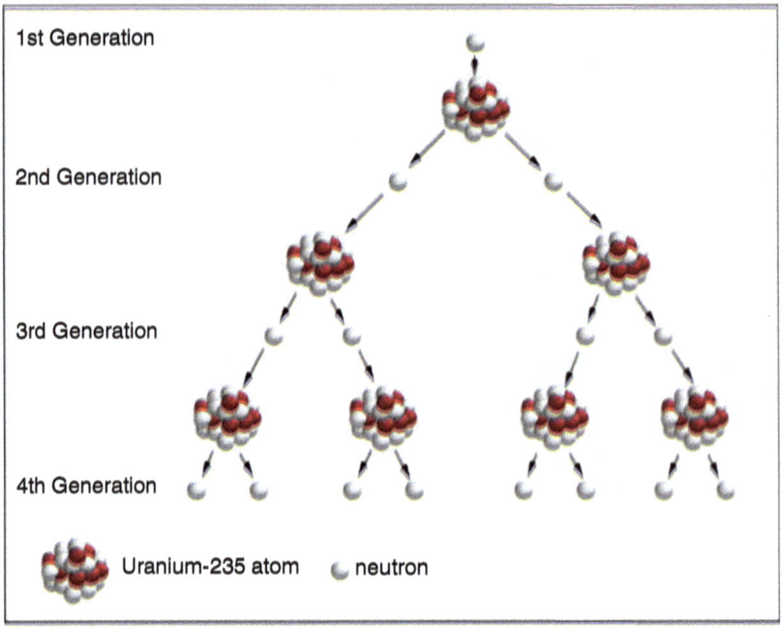

Fig. 3.17 Uranium-235 chain reaction

Table 3.2 Total energy release in U^{235} fission reaction

165 MeV	~Kinetic energy of fission products
7 MeV	~Gamma rays
6 MeV	~Kinetic energy of the neutrons
7 MeV	~Energy from fission products
6 MeV	~Gamma rays from fission products
9 MeV	~Antineutrinos from fission products
Total of 200 MeV	1 MeV (million electron volts) = 1.609×10^{-13} J

g/nucleus, the energy is about 7×10^{17} erg/g. However, the energy release in TNT is 4×10^{10} erg/g or 3.6×10^{16} erg/ton, hence

$$1 \text{ kg of } 25 \approx 2000 \text{ tons of TNT}.$$

Thus, using the chain reaction power $(2)^n$ rule above, we know in 1 kg of Uranium-25 enriched, there are 5×10^{25} nuclei; it would require about $n = 80$ generations of neutrons ($280 \approx 5 \times 10^{25}$) to fissile the whole kilogram.

Moreover, for energy released from each fission reaction, we can write (Table 3.2);

Although two to three neutrons are produced for every fission, not all of these neutrons are available for continuing the fission reaction. If the conditions are such that the neutrons are lost at a faster rate than they are formed by fission, the chain reaction will not be self-sustaining.

At the point where the chain reaction can become self-sustaining, this is referred to as critical mass. In other words, critical mass, from nuclear physics viewpoint, is the minimum amount of a given fissile material necessary to achieve a self-sustaining fission chain reaction under stated conditions. Its size depends on several factors, including the kind of fissile material used, its concentration and purity, and the composition and geometry of the surrounding reaction system.

In an atomic bomb, a mass of fissile material greater than the critical mass must be assembled instantaneously and held together for about a millionth of a second to permit the chain reaction to propagate before the bomb explodes.

The amount of a fissionable material's critical mass depends on several factors: the shape of the material, its composition and density, and the level of purity (e.g., **25**:**1** ratio).

A sphere has the minimum possible surface area for a given mass and hence minimizes the leakage of neutrons. By surrounding the fissionable material with a suitable neutron "reflector," the loss of neutrons can be reduced, and the critical mass can be reduced.

By using a neutron reflector, only about 11 lb (5 kg) of nearly pure or weapon's grade Plutonium-239 or about 33 lb (15 kg) Uranium-235 is needed to achieve critical mass.

In case of nuclear reactor power criticality case, we need to maintain a sustained controlled nuclear reaction for every 2 or 3 neutrons released; only one must be allowed to strike another uranium nucleus. See Fig. 3.18.

If this ratio is less than one, then the reaction will die out; if it is greater than one, it will grow uncontrolled (an atomic explosion). A neutron absorbing the element must be present to control the amount of free neutrons in the reaction space. Most

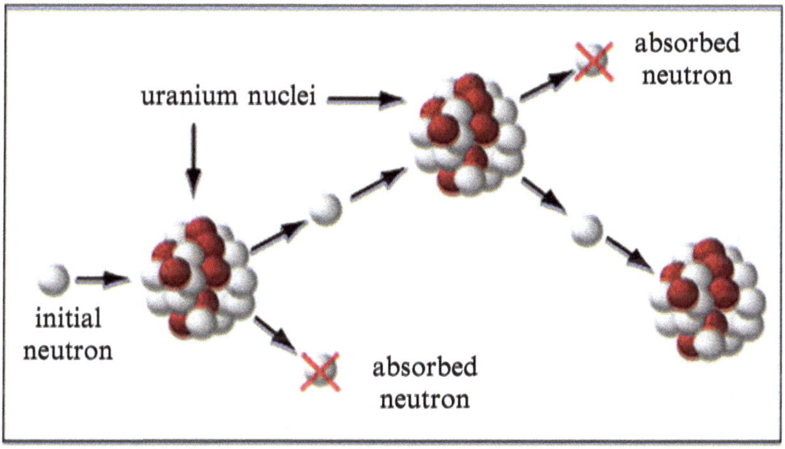

Fig. 3.18 Illustration of controlled nuclear fission

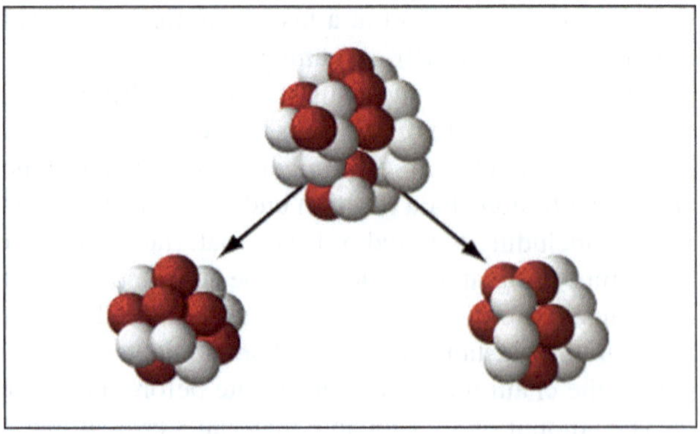

Fig. 3.19 Illustration of spontaneous fission

reactors are controlled by means of control rods that are made of a strongly neutron-absorbent material such as boron or cadmium.

In addition to the need to capture neutrons, the neutrons often have too much kinetic energy. These fast neutrons are slowed using a moderator such as heavy water and ordinary water. Some reactors use graphite as a moderator, but this design has several problems. Once the fast neutrons have been slowed, they are more likely to produce further nuclear fissions or be absorbed by the control rod.

Using Fig. 3.19, the spontaneous nuclear fission rate is the probability per second that a given atom will fission spontaneously, that is, without any external intervention. If a spontaneous fission occurs before the bomb is fully ready, it could fizzle. Plutonium-239 has a very high spontaneous fission rate compared to the spontaneous fission rate of Uranium-235. Scientists had to consider the spontaneous fission rate of each material when designing nuclear weapons.

Detail calculation related to critical mass and associated diffusion theory in particular for sphere shape implosion (i.e., self-similar of second kind) is beyond the scope of this book, but lecture paper by Serber et al. [30] and book by Reed [30] are showing some significant analyses. However, with regard to implosion devices, this neutron initiation window is much smaller; because of the interval during which the bomb is near, optimum criticality is relatively short. Although theoretically it is possible to initiate the fission chain reaction by means of a singular neutron, it is an advantage for an initiator to at least emit several neutrons at the optimum period, because it is possible to capture a singular neutron without causing fission.

A method for initiating the fission chain reaction is to use a continuous neutron emitter: a material, which has a high spontaneous fission rate, or an alpha emitter together with beryllium. Although the neutron production method is stochastic, they are produced with a specific average rate. As a result there will be an uncertainty with regard to the initiation time, which in turn leads to a high degree of variability in the performance of the device, i.e., yield.

An improved version of a continuous/spontaneous neutron emitter is that which can produce a burst of neutrons at an exactly defined time in order to maximize the performance of the device (yield) but at the same time to reduce variability. These so-called internal initiators can be inside the device or external designs, which are positioned outside of the high explosive.

In nuclear reactor dynamics analyses, scientists managed to derive the most accurate description of the neutron behavior in a near-critical reactor for derivation of point kinetics equations (PKE). They deduced a relationship, between power distribution $P(t)$ and reactivity $\rho(t)$ as well as neutron decay β, and they presented the following form of Eq. 3.29, including the term called delay neutron precursors. This precursor is a result of the β-decay of a fission product that leads in some cases to a highly excited product, which can then emit a neutron [35]:

$$\frac{d}{dt}P(t) = \frac{\rho(t) - \beta}{\Lambda}P(t) + \sum_i \lambda_i C_i(t), \tag{3.29}$$

where:

Λ = neutron generation or product time
λ_i = decay constant for delayed neutron precursors of group i
$C_i(t)$ = concentration of delayed neutron precursor of group.

Note that in Eq. 3.29 for the purpose of nuclear reactor dynamics, for reactivity ρ much larger than neutron decay β, i.e., $\rho \gg \beta$, we can ignore the delayed neutron, and as result the precursor equation in point kinetics equation (PKE) goes away. In that case, we are left with the power distribution $P(t)$ with no precursor term as below: [35]

$$\frac{d}{dt}P(t) \approx \frac{\rho(t) - \beta}{\Lambda}P(t). \tag{3.30}$$

The historical reason behind ignorance of the delayed neutron for $\rho \gg \beta$ assumption goes to the nuclear weapon design, where the model was developed and used, hence $\rho \gg \beta$ and rapid transient. The ignorance assumption can be explained by Fuchs–Nordheim for no precursors [35].

Last but not the least, we need to have some idea about the time of reaction for fission to take place in nuclear device criticality. A crude analysis here shows that the released neutron travels at speeds of about 10 Mm s^{-1} or about 3 % the speed of light. The characteristic time for a generation is roughly the time required to cross the diameter of the sphere of fissionable material.

A critical mass of uranium is about the size of a baseball (0.1 m). The time, T, the neutron would take to cross the sphere is

$$\begin{aligned} T &= \frac{0.1\ \text{m}}{1 \times 10^7\ \text{m}\,\text{s}^{-1}} \\ &= 1 \times 10^{-8}\text{s}. \end{aligned}$$

The complete process of a bomb explosion is about 80 times this number or about a microsecond.

This time was informally known as a "shake" ("as fast as the shake of a lamb's tail") by the physicists at Los Alamos.

3.5 Physics of Thermonuclear Explosion

Unlike the Little Boy and Fat Man fission bombs, about which a great deal of information is available in the open literature, virtually no information has been released publicly about the design and construction of thermonuclear weapons. Most of the information that is publicly available is based largely on speculation and supposition, gleaned from open sources and interviews and a few declassified documents.

In contrast to the uranium and plutonium atomic bombs, which depend for their energy on nuclear fission, in which heavy nuclei are broken into lighter components, the hydrogen bomb depends on the energy released during nuclear fusion, in which small light nuclei are fused together to produce larger heavier nuclei. Physicists had known for decades that the fusion process releases an enormous amount of energy.

The principle concept behind thermonuclear explosion and consequently thermonuclear weapons is the energy released from fusion of the two isotopes of hydrogen, namely, deuterium (D) and tritium (T). The chemical reactions of these two isotopes are written below, and because of these reactions that are deviated from hydrogen, the name of hydrogen bomb sticks to thermonuclear weapons.

1. $D + T \rightarrow {}^4He + n(14.1\ MeV) + 17.6\ MeV$
2. $D + D \rightarrow {}^3He + n(2.45\ MeV) + 3.3\ MeV$
3. $D + D \rightarrow T + p + 4.03\ MeV$
4. ${}^3He + D \rightarrow {}^4He + p + 18.35\ MeV$

Out of four chemical reactions of D–T fusion process, the first one is dominating reaction for the design of hydrogen bomb, as well as foundation for the laser-driven fusion in physics of inertial confinement of plasma in generation of future clean energy.

There are other thermonuclear reactions, which can occur. However, the reaction rates are too low in order to be significant. The original idea of thermonuclear weapon was a known fact among the scientists and participants of the Manhattan Project early on and later on was designed, and production was carried at Los Alamos and Lawrence Livermore National Laboratories into its existing thermonuclear weapons.

The first thermonuclear fuel, which was considered for a weapon, was deuterium. This is:

- (Relatively) easy to ignite and burn
- Abundant in nature (in comparison with tritium)
- (Relatively) inexpensive to produce

One other thermonuclear fuel is easier to ignite and this is a mixture of deuterium and tritium. At thermonuclear temperatures, the D + T reaction rate (reaction 1) is two magnitude orders faster than D + D (reaction 2 or 3). However, tritium does not occur in nature and is very expensive to manufacture.

The other method is to produce tritium in situ from other reactions in a functioning weapon. There are two main reactions which produce tritium; reaction 3 (D + D) produces a triton and Lithium-6 and Lithium-7 (7 % and 93 % of natural lithium, respectively) can undergo reactions which produce tritium:

5. $^{6}Li + n \rightarrow T + {}^{4}He + 4.8$ MeV
6. $^{7}Li + n \rightarrow T + {}^{4}He + n - 2.5$ MeV

Because Lithium-6 has a higher neutron cross section than Lithium-7, it is advantageous to enrich the content of Lithium-6 to use as thermonuclear fuel. Because there is always a large excess of deuterium and because of the much higher D + T reaction rate in comparison with the D + D rate, all the tritium produced becomes burned up [29].

As it was stated previously for thermonuclear weapon design, the most efficient way of achieving the required explosion by driving Pu^{239} to its critical mass within a short period of time is implementation of *fusion boosting fission devices.* Figure 3.20 is an illustration of self-similar solution of second kind implemented in the technology of boosting device, in order to increase the efficiency of fission bombs, which results in explosion of thermonuclear bomb, by means of the

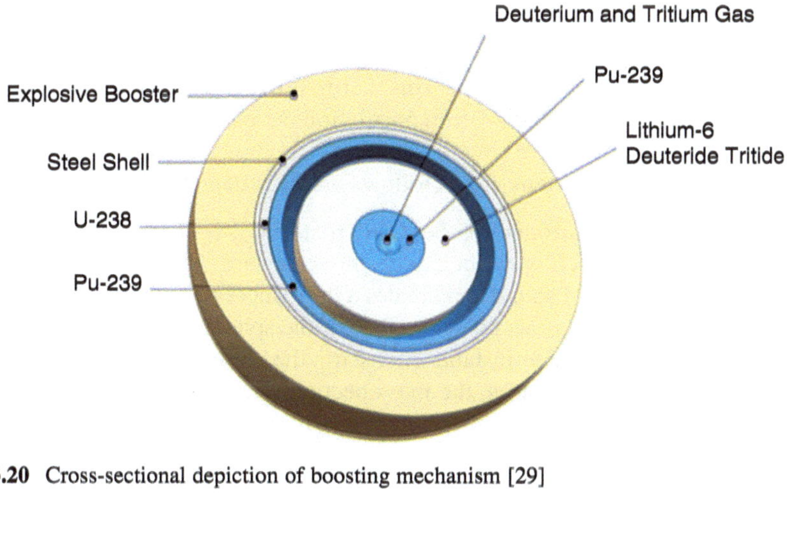

Fig. 3.20 Cross-sectional depiction of boosting mechanism [29]

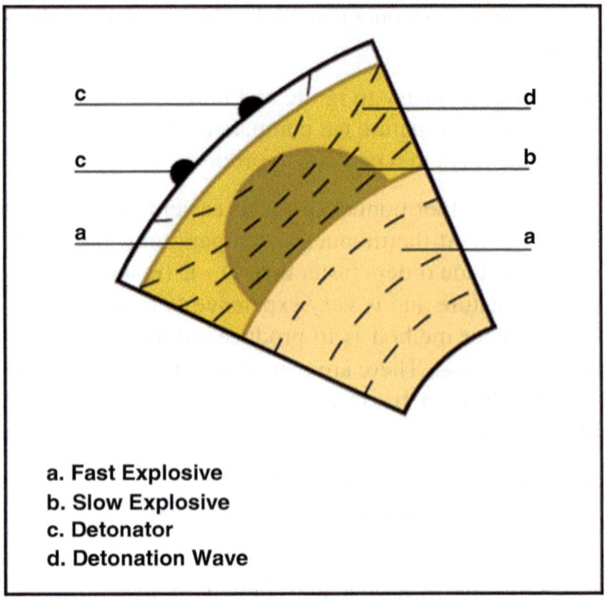

Fig. 3.21 Illustration of detonation sequential steps [47]

introduction of a small amount of deuterium and tritium (typically this contains 2–3 g of tritium) inside the core [29].

As it can be observed in Fig. 3.20, both similarity (i.e., first and second kind) are working hand in hand, where the implosion takes place first and second comes the explosion, and they both power following power laws as follows:

$$\xi = rt^{\alpha} \Rightarrow \begin{cases} \alpha < 0 & \text{Implosion, self-similarity of second kind} \\ \alpha > 0 & \text{Explosion, self-similarities of first kind} \end{cases}. \tag{3.31}$$

In this approach at least at the same time, as the fission chain reaction proceeds, there is a point when the core temperature rises sufficiently for the fusion reaction to begin to occur with a significant rate. This thermonuclear reaction inputs additional neutrons into the core, thus the neutron population increases faster than from fission only [29].

Detonation sequential steps are as follows and it is depicted in Fig. 3.21:

1. The high explosive surrounding the fissile material is ignited.
2. A compressional shock wave begins to move inward. The shock wave moves faster than the speed of sound and creates a large increase in pressure. The shock wave impinges on all points on the surface of the sphere of the fissile material in the bomb core at the same instant. This starts the compression process.
3. As the core density increases, the mass becomes critical and then supercritical (where the chain reactions grows exponentially).

4. Now the initiator is released, producing many neutrons so that many early generations are bypassed.
5. The chain reaction continues until the energy generated inside the bomb becomes so great that the internal pressure due to the energy of the fission fragments exceed the implosion pressure due to the shock wave.
6. As the bomb disassembles, the energy released in the fission process is transferred to the surroundings.

The above steps pretty much follows the Ulam–Teller configuration in version of the Shrimp that is known as the Runt, which was tested at Bikini Island under Operation Castle Romeo in 1954. The Runt used unenriched natural lithium, with 7 % Lithium-6 and 93 % Lithium-7. Again, the yield was much higher than expected—the Runt produced a yield of 11 megatons.

The sequence of the above steps are depicted in the following image sequences as

Figure 3.22a–d clearly is illustrating the physics of similarity both for implosion and explosion for steps of thermonuclear explosion. The most important breakthrough that can be found in open literature was discovered and developed by Sakharov and Yakov Zel'dovich that of using the X-rays from the fission bomb to compress the secondary before fusion ("radiation implosion") in the spring of 1954.

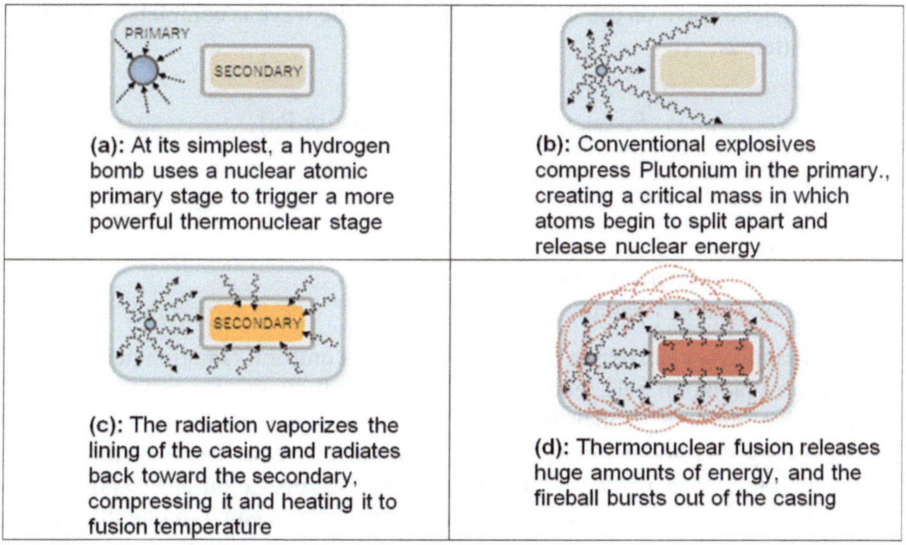

Fig. 3.22 Inside and H-bomb sequence of explosion events (*Source*: "Dark Sun: The making of the hydrogen bomb." By Richard Rhodes). (**a**) At its simplest, a hydrogen bomb uses a nuclear atomic primary stage to trigger a more powerful thermonuclear stage; (**b**) conventional explosives compress plutonium in the primary, creating a critical mass in which atoms begin to split apart and release nuclear energy; (**c**) the radiation vaporizes the lining of the casing and radiates back toward the secondary, compressing it and heating it to fusion temperature; (**d**) thermonuclear fusion releases huge amounts of energy, and the fireball bursts out of the casing

Sakharov's was based on the Teller–Ulam design, which was known in the USSR and was tested in the shot "RDS-37" in November 1955 with a yield of 1.6 Mt. Klaus Fuchs and John Von Neumann in the United States as part of their work on the original "classical super" hydrogen bomb design originally developed the radiation implosion. The physics of such idea is behind the *radiation pressure*, where pressure exerted upon any surface exposed to electromagnetic radiation. Radiation pressure implies an interaction between electromagnetic radiation and bodies of various types, including clouds of particles or gases. The interactions can be absorption, reflection, or some of both (the common case). Bodies also emit radiation and thereby experience a resulting pressure.

The theory behind radiation pressure by absorption using classical electromagnetic waves starts with *Poynting vector*; according to Maxwell's theory of electromagnetism, an electromagnetic wave carries momentum, which can be transferred to a reflecting or absorbing surface hit by the wave. Therefore, the energy flux (intensity) is expressed by Eq. 3.32, whose magnitude is denoted by $|\vec{S}| = S$.

$$\vec{\boldsymbol{S}} = \vec{\boldsymbol{E}} \times \vec{\boldsymbol{H}}. \tag{3.32}$$

S divided by the square of the speed of light in free space is the density of the linear momentum of the electromagnetic field. The time-averaged intensity $< S >$ divided by the speed of light c in free space is the radiation pressure exerted by an electromagnetic wave on the surface of a target, if the wave is completely absorbed and is written as

$$P_{\text{absorb}} = \frac{< S >}{c} = \frac{E_{\text{f}}}{c} \left(\text{N}\,\text{m}^{-2}\,\text{or Pa}\right), \tag{3.33}$$

where:

P = pressure
E_{f} = energy flux (Intensity), in W m^{-2}
c = speed of light in vacuum

Additionally, we can derive the compression in a uniform radiation field. A body in a uniform radiation field (equal intensities from all directions) will experience a compressive pressure. It may be shown by electromagnetic theory, by quantum theory, or by thermodynamics, making no assumptions as to the nature of the radiation (other than isotropy) that the pressure against a surface exposed in a space traversed by radiation uniformly in all directions is equal to one third of the total radiant energy per unit volume within that space. Quantitatively, this can be expressed as follows, per Stefan–Boltzmann black-body radiation and consequently Planck's law for a radiation energy density u (J m^{-3}) as

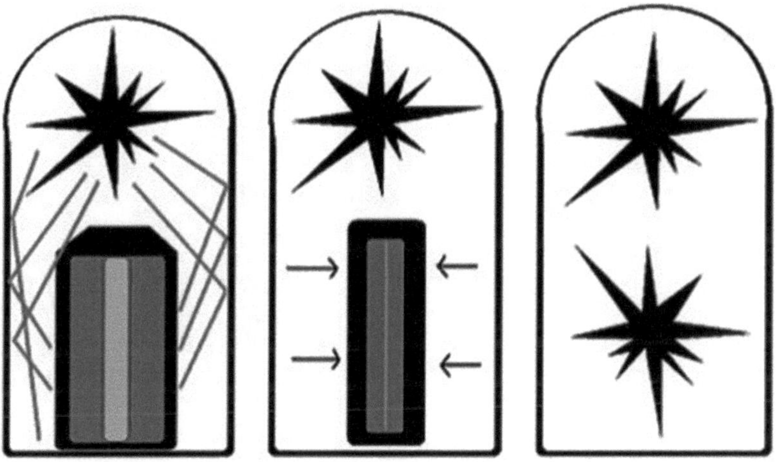

Fig. 3.23 The Ulam–Teller configuration in action

$$P_{\text{compress}} = \frac{u}{3} = \frac{4\sigma}{3c} T^4 \left(\text{N m}^{-2} \,\text{or}\, \text{Pa}\right). \tag{3.34}$$

The second equality holds if we are considering uniform thermal radiation at a temperature T. There σ is the Stefan–Boltzmann constant.

In the above context, we have learned the fundamentals of physics' thermonuclear detonation design process and sequence of process in more details; Fig. 3.23 is expressed here:

Step 1: The exploding primary trigger floods the foam-filled radiation channels at the side of the bomb with X-rays, which are radiated and reflected to the secondary.

Step 2: The outer surface of the secondary pusher ablates or boils away, and the resulting pressure crushes the entire assembly inward, compressing the thermonuclear fuel and imploding the plutonium spark plug.

Step 3: The spark plug detonates, igniting fusion in the lithium fuel. The pusher's momentum helps maintain the fusion reaction for a few microseconds, and finally the Uranium-238 in the pusher undergoes fission from the fast neutrons released by the fusion. The whole process takes a few millionths of a second.

The essentials of a thermonuclear warhead are depicted in Fig. 3.23 as an artistic form as a final possible assembly (Fig. 3.24).

All alone it can be observed that *the problem of implosion* and *explosion at a point*, where both self-similarity first and second kind play a great role, has influence in the thinking of the scientists and physicists involved with the design of the nuclear and thermonuclear weapons.

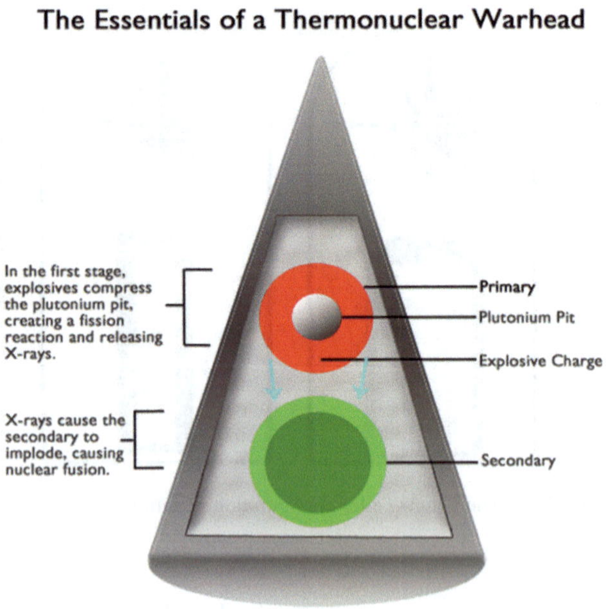

Fig. 3.24 Possible configuration thermonuclear warhead

3.6 Nuclear Isomer and Self-Similar Approaches

Nuclear isomers are long-lived excited states of an atom's nucleus. Some nuclear isomers are naturally occurring, but most of them are produced in artificial nuclear reactions by colliding beams of nuclei and particles. Nuclear isomers decay back to the ground state after some time, which can vary from picoseconds to billions of years, therefore releasing energy in the form of gamma radiation.

If a method could be found to release that energy instantaneously in a gamma ray burst, rather than slowly and at random over time, one would have a new method for controlled high-energy storage and release [36].

The possibility of practical applications of isomers, including for military purposes, has been highlighted in the past few years by the claim that a method for the controlled energy release from one particular isomer, Hafnium178m, had been discovered. However, more experiments that are precise recently showed that this is actually not the case [37].

Consequently, the state of affairs with nuclear isomers is somewhat similar to that of "superheavy elements." Some decades ago, the theoretical prospect for the large-scale synthesis of these elements was uncertain: it is only recently that it has become definitely clear that the production cross sections of stable superheavy elements are too small for any practical applications to be feasible.

In the case of nuclear isomers, the possibility of the existence of a suitable isomer for practical applications (such as a compact energy source for compressing and/or igniting thermonuclear pellets) is entirely open. It could be a matter of pure chance that such an isomer exists and that a method for the controlled release of its energy could be found. In the context above, clearly self-similarity solution for energy release from gamma ray burst can be observed and is only a mathematical solution that deals with continuity equation of motions and the physics theory behind it.

3.7 Pellet Implosion-Driven Fusion Energy and Self-Similar Approaches

Inertial confinement fusion (ICF), in recent years, has raised a lot of interest beyond just the national laboratories in the United States and abroad. ICF aim is toward producing clean energy, using high-energy laser beam or for that matter a particle beam (i.e., the particle beam may consist of heavy or light ion beam) to drive a pellet of two isotopes of hydrogen to fuse and release energy. See the D–T fusion process in Eq. 3.35, where n is neutron and α a particle such as helium ($^4_2$He):

$$\mathrm{D} + \mathrm{T} \rightarrow n(14.06\,\mathrm{MeV}) + \alpha(3.52\,\mathrm{MeV}). \tag{3.35}$$

These two isotopes of hydrogen are known as deuterium (D = ^{2}H) and tritium (T = ^{3}H) as part of fuel, to ignition temperature in order to satisfy the confinement criteria of $\rho r \geq 1$ gram cm^{-2}, where ρ and r are the compressed fuel density and radius pellet, respectively. In order for the confinement criteria also known as Lawson criterion to be satisfied, it needs to take place before occurrence of Rayleigh–Taylor hydrodynamics instability would happen for uniform illumination of the target's surface, namely, pellet of deuterium and tritium.

In direct laser-driven pellet approach, in order to overcome Rayleigh–Taylor instability, we require a large number of laser beams. See Fig. 3.25.

In Fig. 3.25 the schematic of the stages of inertial confinement fusion is done using lasers. The blue arrows represent radiation; orange is blowoff; purple is inwardly transported thermal energy:

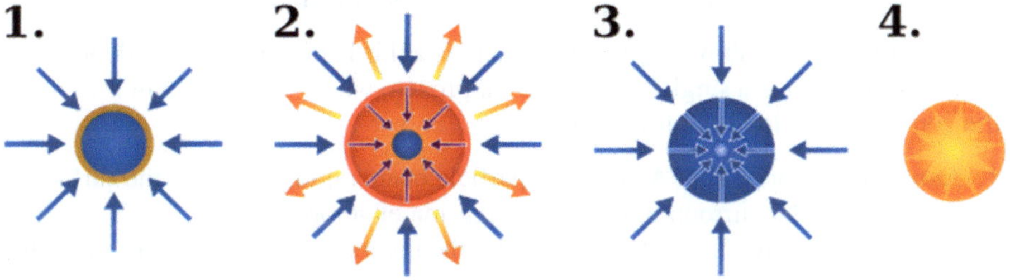

Fig. 3.25 Direct laser-driven compression of a fusion pellet

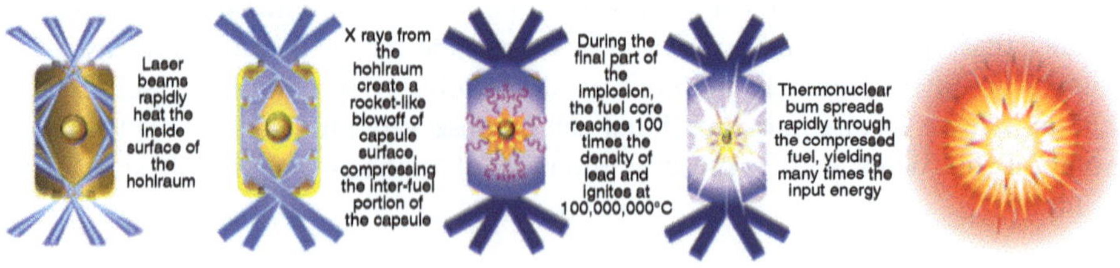

Fig. 3.26 Indirect soft X-ray hohlraum drive compression of fusion of fusion pellet

1. Laser beams or laser-produced X-rays rapidly heat the surface of the fusion target, forming a surrounding plasma envelope.
2. Fuel is compressed by the rocket-like blowoff of the hot surface of the material.
3. During the final part of the capsule implosion, the fuel core reaches 20 times the density of lead and ignites at 100,000,000 °C.
4. Thermonuclear burn spreads rapidly through the compressed fuel, yielding many times the input energy.

In case of indirect illuminating target approach, the laser light is converted into soft X-ray, which is trapped inside a hohlraum chamber surrounding the fusion fuel irradiating it uniformly. In this approach, in order to archive fusion inertial confinement, the energy source that drives the ablation and compression, as it was stated, is soft X-ray ration. This is produced by the conversion of a nonthermal, directed energy source, such as lasers or ion beams, into thermal radiation inside a high-opacity enclosure that is referred to as a hohlraum. See Fig. 3.26.

In Fig. 3.26 the schematic of the stages of inertial confinement fusion using lasers driving the pellet, the compression proceeds along several steps from left to right as:

1. Laser illumination: Laser beam rapidly heat the inside surface of the hohlraum.
2. Indirect drive illumination: The walls of the hohlraum create an inverse rocket effect from the blowoff of the fusion pellet surface, compressing the inner fuel portion of the pellet.
3. Fuel pellet compression: During the final part of the implosion process, the fuel core reaches a high density and temperature.
4. Fuel ignition and burn: The thermonuclear burn propagates through the compressed fusion fuel amplifying the input energy in fusion fuel burn.

In addition to the above approaches, there is a third approach, as it is depicted in Fig. 3.27, and that is a single-beam direct approach, where a single beam is used for the compression along the following steps:

1. Atmospheric formation: A laser or a particle beam rapidly heats up the surface of the fusion pellet surrounding it with a plasma envelope.
2. Compression: The fuel is compressed by the inverse rocket blowoff of the pellet surface imploding it inward.

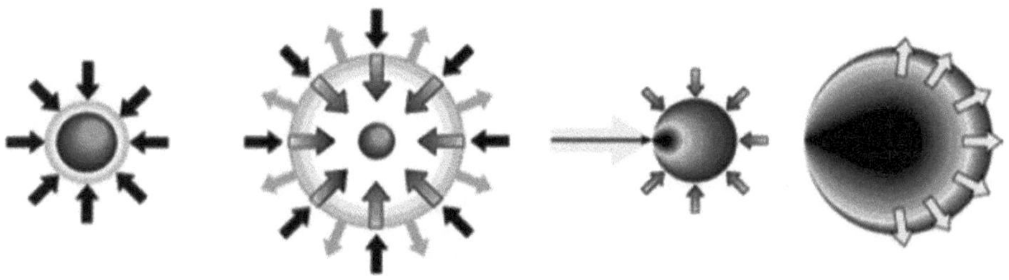

Fig. 3.27 Single-beam igniter concept for fusion pellets

3. Beam fuel ignition: At the instant of maximum compression, a short high-intensity pulse ignites the compressed core. An intensity of 10^{19} [Watts cm^{-2}] is contemplated with a pulse duration of 1–10 μs.
4. Burn phase: The thermonuclear burn propagates through the compressed fusion fuel yielding several times the driver input energy.

In either approaches above, Lawson criterion for the simple case of physics of inertial confinement fusion (ICF) can easily be calculated as follows.

The Lawson criterion applies to inertial confinement fusion (ICF) as well as to the magnetic confinement fusion (MCF) but is more usefully expressed in a different form. A good approximation for the inertial confinement time τ_E is the time that it takes an ion to travel over a distance r at its thermal speed v_{Thermal}:

$$v_{\text{Thermal}} = \sqrt{\frac{k_B T}{m_i}}, \tag{3.36}$$

where:

k_B = Boltzmann constant
m_i = mean ionic mass
T = temperature

Equation 3.36 is derived from kinetic energy theory and gas pressure relationship.

The inertial confinement time τ_E can thus be approximated as

$$\tau_E \approx \frac{r}{v_{\text{Thermal}}}. \tag{3.37}$$

Substituting Eq. 3.36 into Eq. 3.37 results in

$$\tau_{\mathrm{E}} \approx \frac{r}{v_{\mathrm{Thermal}}} = \frac{r}{\sqrt{\frac{k_{\mathrm{B}}T}{m_i}}} = r \cdot \sqrt{\frac{m_i}{k_{\mathrm{B}}T}} \quad (3.38)$$

However, the Lawson criterion requires that fusion heating fE_{ch} exceeds the power losses P_{loss} as written below:

$$fE_{\mathrm{ch}} \geq P_{\mathrm{loss}}. \quad (3.39)$$

In this equation the *volume rate* f is the reactions per volume time of fusion reaction and is written as

$$f = n_{\mathrm{Deuterium}} n_{\mathrm{Tritium}} <\sigma v>= \frac{1}{4} n^2 <\sigma v> . \quad (3.40)$$

Moreover, E_{ch} is the energy of the charged fusion products, and in case of deuterium–tritium reaction, it is equal to 3.5 MeV.

In addition, power loss density P_{loss} is the rate of emery loss per unit volume and is written as

$$P_{\mathrm{loss}} = \frac{W}{\tau_{\mathrm{E}}}, \quad (3.41)$$

where W is the energy density or energy per unit volume and is given by

$$W = 3nk_{\mathrm{B}}T. \quad (3.42)$$

In all the above equations, the variables that are used are defined as below:

k_{B} = Boltzmann constant
n = particle density
$n_{\mathrm{Deuterium}}$ = deuterium particle density
n_{Tritium} = tritium particle density
τ_{E} = confinement time that measures the rate at which a system loses energy to its surrounding environment
σ = fusin cross section
v = relative velocity
$<\sigma v>$ = average over the Maxwellian velocity distribution at temperature T
T = temperature

Now substituting for all the quantities in Eq. 3.39, the result is written as

$$n\tau_{\mathrm{E}} \geq \frac{12}{E_{\mathrm{ch}}} \frac{k_{\mathrm{B}}T}{<\sigma v>} \equiv L. \tag{3.43}$$

Equation 3.43 is known as the Lawson criterion, and for the deuterium and tritium reaction, it is at least $n\tau_{\mathrm{E}} \geq 1.5 \times 10^{20}$ s m^{-3}, where the minimum of the product occurs near $T = 25$ keV. The quantity $T/ < \sigma v >$ is a function of temperature with an absolute minimum. Replacing the function with its minimum value provides an absolute lower limit for the product $n\tau_{\mathrm{E}}$.

Substituting Eq. 3.43 into Eq. 3.38, we obtain

$$n\tau_{\mathrm{E}} \approx n \cdot r \cdot \sqrt{\frac{m_i}{k_{\mathrm{B}}T} \geq \frac{12}{E_{\mathrm{ch}}} \frac{k_{\mathrm{B}}T}{<\sigma v>}} \tag{3.44}$$

or

$$n \cdot r \geq \frac{12}{E_{\mathrm{ch}}} \cdot \frac{(k_{\mathrm{B}}T)^{3/2}}{<\sigma v> \cdot m_i^{1/2}}. \tag{3.45}$$

Equation 3.45 could be approximated to the following form as

$$n \cdot r \geq \frac{(k_{\mathrm{B}}T)^{3/2}}{<\sigma v>}. \tag{3.46}$$

This product must be greater than a value to the minimum of $T^{3/2}/ <\sigma v>$. The same requirement is traditionally expressed in terms of mass density $\rho = <nm_i>$ as

$$\rho r \geq 1 \text{ g/cm}^2. \tag{3.47}$$

Satisfaction of this criterion at the density of solid deuterium–tritium (0.2 g cm^{-3}) would require a laser pulse of implausibly large energy. Assuming the energy required scales with the mass of the fusion plasma $(E_{\mathrm{Laser}} \approx \rho r^3 \approx \rho^{-2})$, compressing the fuel to 10^3 or 10^4 times solid density would reduce the energy required by a factor of 10^6 or 10^8, bringing it into a realistic range. With a compression by 10^3, the compressed density will be 200 g cm^{-3}, and the compressed radius can be as small as 0.05 mm. The radius of the fuel before compression would be 0.5 mm. The initial pellet will be perhaps twice as large, since most of the mass will be ablated during the compression.

The fusion power density is a good figure of merit to determine the optimum temperature for magnetic confinement, but for inertial confinement, the fractional burnup of the fuel is probably more useful. The burnup should be proportional to the specific reaction rate ($n^2 <\sigma v>$) times the confinement time (which scales as $T^{-1/2}$) divided by the particle density n:

Fig. 3.28 Depiction of the reduced self-heating of the hot spot area from prematurely escaping particles

$$\text{burn-up fraction} \quad \Rightarrow \begin{cases} \propto & n^2 < \sigma v > T^{-1/2}/n \\ \propto & (nT) < \sigma v > /T^{3/2} \end{cases}. \tag{3.48}$$

Thus the optimum temperature for inertial confinement fusion maximizes $<\sigma v> /T^{3/2}$, which is slightly higher than the optimum temperature for magnetic confinement.

Note that part of key issues, for laser to drive pellet of micro-balloon containing deuterium and tritium to achieve fusion, is a symmetrical homogenous compression, which means targeting for perfectly spherical implosions and explosions. However, in reality, this ideal situation never will take place to its perfection and as a result has a number of physics problem consequences and they are:

- Instabilities and mixing
  - Rayleigh–Taylor unstable compression [12]
  - Break of symmetry destroys confinement
- How to improve energy coupling into target (i.e., pellet of D–T), which requires the conversion of kinetic energy from the implosion into internal energy of the fuel that is not perfect. Additionally, we need to prevent the reducing of the maximum compression.
- Severe perturbing of a spherical homogeneous and symmetric implosion can result in small-scale turbulences and even breakup of the target shell.
- The hot spot area at the ablation surface is increased or has a large surface due to the perturbed structure, which leads to reduction of ignition temperature to achieve fusion reaction in the corona of the pellet, and it causes the α-particle

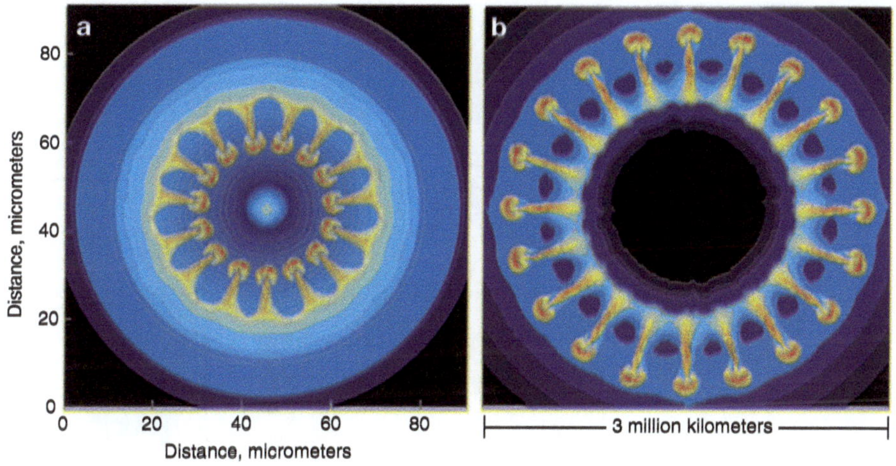

Fig. 3.29 Striking similarities exist between hydrodynamic instabilities in (**a**) inertial confinement fusion capsule implosions; (**b**) core-collapse supernova explosions

created in Eq. 3.35 to escape the hot spot area. This also lowers the self-heating. See Figs. 3.28 and 3.29 below.

- Finally, what is the best material for the first wall of pellet of D–T as a target?

In summary, the Rayleigh–Taylor instabilities occur when a lower-density fluid such as, for example, oil, underlies a higher density fluid such as water. In inertial confinement where the implosion and explosion process takes place in a sequence, the higher density fluid is the pellet surface, and the lower-density fluid is the plasma surrounding it and compressing the pellet through the inverse rocket action (i.e., inertial) of the implosion process.

In all approaches stated above for the inertial confinement via laser or particle beams imploding and exploding, target pellet in a symmetrical and homogeneous mode is mainly influenced by Rayleigh–Taylor (RT) instabilities at the ablation surface.

As we stated, the impact and effect of the Rayleigh–Taylor (RT) instabilities are because they initially grow exponentially so that even very small and insignificant disturbances can grow to a size that has adverse effect on the entire compression in homogeneous and symmetrical mode as it is observed in the Fig. 3.30 illustration. In this illustration, the major instability is again because of heavy material pushed on low density one.

This instability always occurs, since the laser or particle beam as driver of deuterium–tritium pellet is never 100 % homogeneous and symmetric; consequently, the Rayleigh–Taylor instability always is growing.

The growth rate of the Rayleigh–Taylor instability can be measured in a wavelength range not previously accessible, and it is a very important factor that one needs to pay attention to it during the implosion and explosion of pellet.

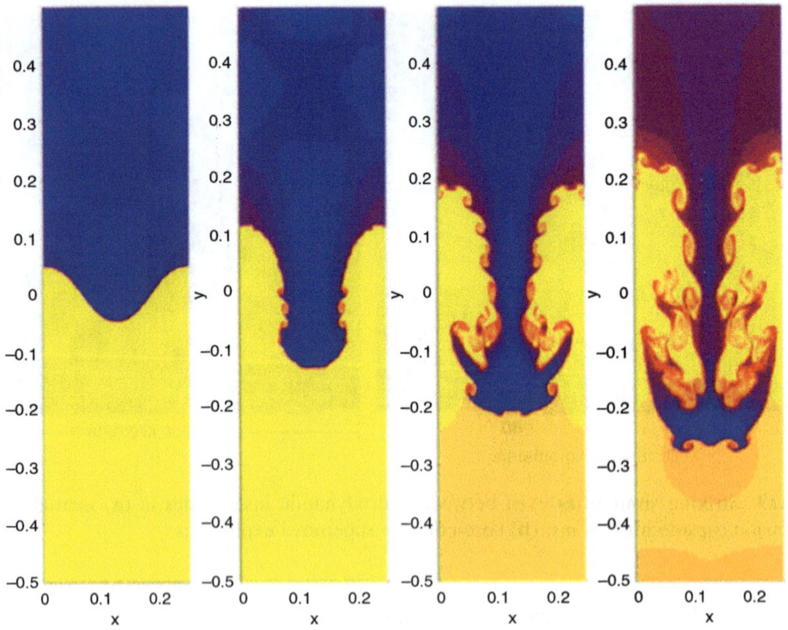

Fig. 3.30 Growth of Rayleigh–Taylor instability illustration

Fig. 3.31 Growth of Rayleigh–Taylor instabilities during pellet implosion

Moreover, it is important for the purpose of delivering energy to the corona of pellet as symmetric and homogenous as possible, before the plasma frequency generated at the ablation surface reaches beam wavelength frequency as driver. See Fig. 3.31.

Thus, in conclusion, the fusion targets can be illuminated with the energy of different drivers. The primary efforts in inertial confinement exist in the United States, France, and Japan.

3.7.1 *Linear Stability of Self-Similar Flow in D–T Pellet Implosion*

As part of the effort for laser or particle beam to drive D–T pellet for release of energy via fusion reaction, it is the study of linear stability of self-similar flow where we need to take under consideration the analyses of imploding cylindrical and spherical shocks in what is known as Chester–Chisnell–Whitham (C–C–W) approximation method [20–22].

The linearized Chester–Chisnell–Whitham (C–C–W) equations describing the evolution of an arbitrary perturbation about an imploding shock wave in an ideal fluid are solved exactly in the strong shock limit for a density profile $\rho(r) \sim Ar^{-q}$, where A and q are constants. All modes are found to be relatively unstable (i.e., the ratio of perturbation amplitude to shock radius diverges as the latter goes to zero), provided that q is not too large. The nonlinear C–C–W equations are solved numerically for both moderate and strong shocks.

The small amplitude limit agrees with the analytical results, but some forms of perturbation, which are stable at small amplitude, become unstable in the nonlinear regime. The results are related to the problem of pellet compression in experiments on inertial confinement fusion. This study extensively is done by Gardner et al. [26].

One of the critical limitations to achieving high compression in a spherical implosion is the degree of symmetry that can be maintained. This in turn has important implications for target fabrication techniques and for laser or other driver designs, since it establishes the tolerances required in the symmetry of these components [26].

An important issue for understanding imploding systems is the stability of a converging shock wave. This shock wave might be used, for instance, not only to compress the fuel but also to provide the heating required to create a central ignition region. The final temperature achieved will depend on how nearly spherical the shock wave remains during the collapse process and the shape of the shock at the time of reflection [26].

A certain inherent stability of a shock wave results from the well-known fact that a shock wave with a smaller radius of curvature advances faster than one with a larger radius of curvature. Thus, the part of a perturbed shock front that initially lags behind will accelerate more rapidly due to its smaller radius of curvature and so will tend to catch up with the remainder of the shock wave. However, the perturbation may be unable to overtake the main shock, which is accelerating because of convergence, or it may be overdriven, i.e., the perturbation may overshoot the stable position [26].

In order to discuss stability, it is necessary to define what is meant by stable (or unstable) behavior. The usual definition of stability in terms of growing or decaying mode amplitude does not adequately describe the situation in imploding systems. For example, the amplitude may not tend to zero as fast as the average radius, or the collapse time may be of the order of the period of oscillation of the mode. A more meaningful number for small amplitude perturbations is the rate of growth (or decay) of the relative perturbation amplitude, i.e., the ratio of the perturbation amplitude to the radius of the zero-order symmetric collapse solution (Bernstein and Book, 1978) [38].

We strongly encourage our reader to refer to the article published by Gardner et al. [26].

3.8 Plasma Physics and Particle-in-Cell Solution (PIC)

Particle-in-cell that is known as PIC is a method which refers to a technique used to solve a certain class of partial differential equations. In this method, individual particles (or fluid elements) in a Lagrangian frame are tracked in continuous phase space, whereas moments of the distribution such as densities and currents are computed simultaneously on Eulerian (stationary) mesh points.

The method gained popularity for plasma simulation in the late 1950s and early 1960s by Buneman, Dawson, Hockney, Birdsall, Morse, and others. In plasma physics applications, the method amounts to following the trajectories of charged particles in self-consistent electromagnetic (or electrostatic) fields computed on a fixed mesh. To develop point designs for fast ignition, laser interaction simulations with peak densities at, or approaching, solid density are required. Simulating high-density plasmas places severe constraints on the use of standard particle-in-cell (PIC) techniques to solve Maxwell's equations and the particle equations of motion, limiting the spatial and temporal scales which can be easily modeled.

In order to deal with high-density plasma, we need to encounter the problem of solving magnetohydrodynamics (MHD) equations that are required in the study of either magnetic confinement fusion (MCF) or inertial confinement fusion (ICF). Therefore, a method for solving hydrodynamic problems involving large distortions and compression of the fluid in several space dimensions is required.

The particle-in-cell method for hydrodynamic calculations involving the conservation equations of motion is one way to deal with this problem. The calculation procedure introduces finite difference approximations to the partial differential equation; the solution in practice could be carried out by means of high-speed computers, when there is no exact solution existing for such complex nonlinear partial differential equation, while we can use similarity approach to reduce them to ordinary differential equations.

Additionally, realistic studies of the complicated dynamics of compressible fluids are being made possible by recent development of powerful and high-speed computing machines and hardware. Methods are well known for treating problems dependent upon one space coordinate only in one-dimensional problems either in Lagrangian (L), Eulerian (E), or combination of both known as the arbitrary Lagrangian and Eulerian (ALE). For those dependent upon two or more space coordinates, two-dimensional problems, etc., several methods of treatment could possibly be used in some restricted classes of problems.

Particle-in-cell suggests such procedure to the system of equations, which are subject to initial boundary conditions, and they could be the continuity equations of motion of particles of interest in fluid.

3.9 Similarity Solutions for Partial and Differential Equations

Systems of differential equations occur often in many theoretical and applied areas. In many cases, exact solutions are required as numerical methods are not appropriate or applicable. Indeed, exact solutions of systems of partial differential equations arising in fluid dynamics, continuum mechanics, and general relativity are of considerable value for the light they shed into extreme cases, which are not susceptible to numerical treatments.

Normally the symmetrical problems are the simplest ones to solve. Less obvious but equally real is the algebraic symmetry of ordinary and partial differential equations, which, if present, can facilitate the solution of the equations just as geometrical symmetry can facilitate the solution of geometrical problems. On the other hand, what the algebraic symmetry means can be described according to usage as long as an operation on it leaves it looking the same. For example, rotating an equilateral triangle by 120° around its centroid does not change its appearance. This means that symmetry of the triangle by saying it is *invariant* to rotation of 120° around its centroid. It is invariant to other transformations as well; if we reflect it in about one of its altitudes, the image looks the same as the object. An even more symmetric shape than the above triangle is the circle, and it is invariant to all rotations around its centroid and to reflection in any diameter.

Both ordinary and partial differential equations are sometimes invariant to groups of algebraic transformations, and these algebraic invariance, like the geometric ones mentioned above, are also symmetries.

One important source of exact solutions to differential equations is the application of the group theoretic method of Lie. Such solutions found by Lie's method are called invariant solutions. He showed how to use knowledge of the transformation group:

1. To construct an integrating factor for first-order ordinary differential equations
2. To reduce second-order ordinary differential equations to first order by a change of variables

These two results are the more important because *they do not depend on the equation as being linear*.

When Boltzmann used the algebraic symmetry of the partial differential diffusion equation to present his study of diffusion with a concentration-dependent diffusion coefficient, he never made explicit mention of transformation groups or symmetry, but his method was using the symmetry to find special solution of the partial differential equation by solving a related *ordinary* differential equation. See Example 1, Sect. 4.2, and Chap. 4 of the book. Later on the American mathematician Garrett Birkhoff [39] was the first one to recognize that Boltzmann's procedure depended on the algebraic symmetry of the diffusion equation and could be generalized to other partial differential equations, *including nonlinear one* [40]. Using the algebraic symmetry of the partial differential equation, he showed how solution could be found by solving related ordinary differential equation, which such solutions are called similarity solutions (see Chap. 4 for more details).

Essential to this approach is the need to solve overdetermined systems of determining equations, which consist of coupled, linear, homogeneous, partial differential equations. Typically, such systems vary between ten to several hundred equations. Clearly in the case of sets of equations consisting of about 100 equations or more, the prospect of finding solutions to such systems with just pencil and paper would certainly be quite daunting.

DESOLV, which runs under Maple, attempts to automate as much as possible the process of determining these invariant solutions. The program has modular structure and not only uses basic features of Maple but also has independently built-in routines to augment or assist. DESOLV is simply a solver, which attempts to find the solutions of a linear or nonlinear system of overdetermined differential equations of polynomial type. The implementation of DESOLV is based on a heuristic procedure, which does not always promise to produce complete solutions of a system. If the heuristic procedure fails to simplify a system completely, it will transform the system of unsolved equations into a standard form. A best, published paper is given by Vu, Butcher, and Carminati [41], where they presented similarity solutions of partial differential equations using DESOLV.

3.10 Dimensional Analysis and Intermediate Asymptotic

Among the theoretical method for solving many problems of applied mathematics, physics, fluid mechanics, and any related technologies both numerically and analytically using an approach known as *asymptotical method* needs some special attention. In this chapter the main aim is to introduce the reader to the mathematical methods of asymptotic and perturbation theory and to be able to review later on the most important methods of singular perturbations within the scope of application of differential equations.

Asymptotic method is a method by which very often results in discovering the essential characteristics of analyzed processes. The results very often serve as the verification and are used in tests, which leads to obtaining more effective algorithms of numerical evaluation. The obtained results turned out to be quantities for some ranges of parameter changes as well as changes of reference inputs of the processes under consideration.

The dynamics of the development of perturbation theory and asymptotic methods will allow certain nonlinear differential equation to be solved in a very satisfactory manner along with associated boundary and initial conditions. Let us consider how the solution of differential equation depends on parameter ε coordinates and time, and let us assume that, for the small ε, we know the formula describing the "deviation" from the known boundary solution obtained for $\varepsilon = 0$ which is so-called generating solution. Since both the analyzed differential equation and boundary conditions depend on ε, and since the deviation mentioned above is determined by the application of analytical transformations, we will assume that it depends analytically also on ε. Due to that, the perturbance will be always searched in the form of asymptotic solutions with regarding the small parameter ε.

The simplest example of *perturbation theory and analyses* is the classical direct application of *the method of small parameters*. However, as it turns out in most of the cases of irregular or singular problem, the results obtained from the classical approach are not satisfactory and may lead to false conclusions [1]. The desire for solving this essential problem through decades has led many physicists and mathematicians to study various asymptotic methods of singular perturbations, which allows taking into account nonlinearity of the analyzed problems. According to Van Dyke [42], the characteristic feature of the problem of singular excitations is the fact that none of the asymptotic series is uniformly adapted for application in the completely researched range of solutions. Moreover, the problem of singular perturbations occurs also when not only the first approximation but also approximations of higher order have the mentioned characteristic [42].

Asymptotic methods may be useful for investigating even more complex mathematical problems occurring during mathematical modeling and scaling of some affecting wave processes in fluid and gas dynamics.

In Sect. 4.7 of Chap. 4, we will take under consideration nonlinear problems for which the exact solution is not tenable; one may look for asymptotic solutions for a large time or distance.

At the beginning of Chap. 1, we introduced the definition of spherical and cylindrical-like piston wave, such as Taylor explosion [3, 4] and Guderley implosion [8] problems. The induced shock waves in both cases were outlined, and it was illustrated how an application of elongated coordinates and renormalization methods yields highly accurate results. We also briefly showed how an application of the methods of renormalization, characteristic, and multiple scales allowed for overcoming the singularities to achieve the uniformly suitable solutions. In order to look at self-similarity and intermediate asymptotics, we refer to Barenblatt [38] book Chap. 2 as well as the paper published by Barenblatt and Zel'dovich [43], which deals with "Self-Similar Solutions as Intermediate Asymptotic." Per their statements, a certain phenomenon that is time independent will be at *steady state*, and dealing with this type of PDE or ODE is by far easier since there is no need to trace its evolution in time. Moreover, the other significant phenomenon that we can take under consideration in solving somewhat complex physics and engineering is known as *self-similar*. What that means is the case that the *spatial distributions of the characteristic of the phenomenon* such as flow velocity, stress, electric current, etc. They define a function of the form as $U(\vec{r}, t)$ and *vary with time while remaining geometrically similar* while there exist time-dependent scales $U_0(t)$ and $r_0(t)$ such that measured in these scales the phenomenon becomes time independent such as

$$U(\vec{r}, t) = U_0(t) f\left(\frac{\vec{r}}{r_0(t)}\right). \tag{3.49}$$

Partial differential equations of physics and engineering science along with their applications that require to be dealt with via self-similar approach, dimensional analysis, and mathematical modeling can be seen by many fluid and gas dynamics books as well as few examples that were presented in Chaps. 4 and 5 of this book. One of the most astonishing examples is Taylor explosion [1] and Guderley implosion [8] problems that are demonstrated in Chap. 1 where the distributions of every property $U(\vec{r}, t)$ of Eq. 3.49 is reduced to the following self-similar form that can be represented as follows in the case of Taylor strong explosion:

$$u = u_0(t) f\left(\frac{r}{r_f(t)}, \gamma\right), \tag{3.50}$$

where γ is a constant adiabatic index, $r_f(t) = C(\gamma)\left(Et^2/\rho_0\right)^{1/5}$ and $u_0(t) = \rho_0$ for the density, $u_0(t) = \rho_0^{3/5} E^{2/5} t^{-6/5}$ for the pressure, and $u_0(t) = \left(Et^{-3}/\rho_0\right)^{1/5}$ for the velocity [1].

As one can see, self-similar solutions are encountered in many branches of mathematical physics. Searching for a self-similar solution has always been a challenge for the scientist, where the basic point is that in most cases, it allows reduction of nonlinear partial differential equations to be the one involving ordinary differential equations in a problem of interest. See Chap. 4 for details and group theory discussing this situation. Although today's approaches with power of computer mainly are numerical rather than analytical solutions.

Barenblatt [7] also expresses that "self-similar solutions are always solution to limiting, 'idealize', problems for which the parameters having the same dimensions as the independent variables are equal to zero or infinity. Thus in the very-intense-explosion problem the initial energy was assumed to be concentrated in a point, the explosion was assumed to be instantaneous and the initial air pressure in the ambient atmosphere to be negligibly small."

Although at the beginning of the discovery of self-similar method approaches every solution of this kind were treated by most scientist as they were limited to "exact special solutions" to very special problems, gradually researchers working in this field realized that these solutions were actually of much broader significance than that of isolated cases. Therefore with rise of self-similar solution and its description of behavior physical system under some special conditions but leads to solution of broader classes that induced from known method as *intermediate asymptotics* behavior of these types of problems. A behavior in regions where these solutions have ceased to depend on the fine details of the initial or boundary conditions, but where the system still far from its final equilibrium state both large in time and distance from this final point. This is a common condition for any asymptotic solution of any partial differential or ordinary differential equations, and it greatly increases the significance of self-similar method to be used.

For exact definition of *intermediate asymptotics*, we refer to Barenblatt's [44] description and bear in mind that asymptotic is an approximate representation of a function that is valid in a certain range of independent variables within that function. He assumes that in the phenomenon under consideration, there exist two values of an independent variable x, x_1 and x_2, having widely different magnitudes and in a sense $x \gg x_1$ while $x \ll x_2$ so we can show

$$x_1 \ll x_2. \tag{3.51}$$

Then the asymptotic representation of certain properties of the phenomenon in the range is given by

$$x_1 \ll x \ll x_2. \tag{3.52}$$

Corresponding to values of the independent variable x that are large in comparison with the first scale x_1 but small in comparison with the second scale x_2 is called the *intermediate asymptotics*. To be more precise, if in a problem we face two widely different scales x_1 and x_2 in the values of an independent variable x, then we call the *intermediate asymptotics* an asymptotic representation for $\frac{x}{x_1} \to \infty$ but meanwhile $\frac{x}{x_2} \to 0$.

In addition, one should bear in mind that dealing with asymptotic method while constructing approximate solutions to algebraic, differential, and integral equations and their system, as well as during the estimation of various integrals, *asymptotic series* are widely applied with respect either to a parameter or to an independent variable [45]. The mentioned expansion into power series are usually constructed using both the Taylor and Maclaurin series or using other special tools.

It also worth to mention that in asymptotic approaches in order to solve mathematics, physics, and engineering, the perturbation techniques can be divided into two main groups known as:

1. Regular problems that are associated with the use of a classical method of small parameter, i.e., they allow for finding an asymptotic decomposition uniform in the whole domain.
2. Irregular also known as singular problems cannot be solved through a classical small parameter method only in some bounded domain of a sought function, and those subspaces are called *singular subspaces of asymptotic decompositions*.

Some introductory problems related to nonuniformity of an asymptotic decomposition can be found in Sect. 2.2.4 of Awrejcewicz and Krysko's book [45]. Although now these days, there are many various methods of the theory of singular perturbations allowing for the construction of uniformly applicable asymptotic decomposition in various branches of applied mathematics, physics, and other related sciences.

3.11 Asymptotic Analysis and Singular Perturbation Theory

In this section, we describe the aims of perturbation theory in general terms and give some simple illustrative examples of perturbation problems, and for that some lectures are given on this subject by Professor John K. Hunter [46]. To explain the *Perturbation Theory*, we consider a problem of the following form:

$$P^{\varepsilon}(x) = 0 \tag{3.53}$$

depending on a small, real-valued parameter ε that simplifies in some way when $\varepsilon = 0$ (e.g., it is linear or exactly solvable). The aim of perturbation theory is to determine the behavior of the solution $x = x^{\varepsilon}$ of Eq. 3.53 as $\varepsilon \to 0$. The use of a small parameter here is simply for definiteness, for example, a problem depending on a large parameter ω can be rewritten as one depending on a small parameter $\varepsilon = 1/\omega$.

The focus of these notes is on perturbation problems involving differential equations, but perturbation theory and asymptotic analysis apply to a broad class of problems. In some cases, we may have an explicit expression for x^{ε}, such as an integral representation, and want to obtain its behavior in the limit $\varepsilon \to 0$.

The first goal of perturbation theory is to construct a formal asymptotic solution of Eq. 3.5 that satisfies the equation up to small error. For example, for each $N \in \mathbb{N}$ we may be able to find an asymptotic solution x_N^{ε} such that

$$P^{\varepsilon}\left(x_N^{\varepsilon}\right) = O\left(\varepsilon^{N+1}\right), \tag{3.54}$$

where $O(\varepsilon^n)$ denotes a term of the order ε^n. This notation O and o provides a precise mathematical formation of ideas that correspond—roughly—to the "same order of magnitude" and "smaller order of magnitude." We state the definitions for the asymptotic behavior of a real-valued function $f(x)$ as $x \to 0$, where x is a real parameter. With obvious modifications, similar definitions apply to asymptotic behavior in the limits $x \to 0^+, x \to x_0$, and $x \to \infty$ to complex or integer parameters and other cases. Let us look at few examples.

Examples A few simple examples are:

(a) $\sin\frac{1}{x} = O(1)$ as $x \to 0$.
(b) It is not true that $1 = O(\sin 1/x)$ as $x \to 0$, because $\sin 1/x$ vanishes in every neighborhood of $x = 0$.
(c) $x^3 = o(x^2)$ as $x \to 0$ and $x^2 = o(x^3)$ as $x \to \infty$.
(d) $x = o(\log x)$ as $x \to 0^+$ and $\log x = o(x)$ as $x \to \infty$.
(e) $\sin x \tilde{} x$ as $x \to 0$.
(f) $e^{-1/x} = o(x^n)$ as $x \to 0^+$ for any $n \in \mathbb{N}$.

The o and O notations are not quantities without estimates for the constants C, δ, and r appearing in the definitions.

3.12 Regular and Singular Perturbation Problems

It is useful to make an imprecise distinction between regular perturbation problems and singular perturbation problems. A regular perturbation problem is one for which the perturbed problem for small, nonzero values of ε is qualitatively the same as the unperturbed problem for $\varepsilon = 0$. One typically obtains a convergent expansion of the solution with respect to ε, consisting of the unperturbed solution and higher-order corrections. A singular perturbation problem is one for which the perturbed problem is qualitatively different from the unperturbed problem. One typically obtains an asymptotic, but possibly divergent, expansion of the solution, which depends singularly on the parameter ε.

Although singular perturbation problems may appear as typical, they are the most interesting problems to study because they allow one to understand qualitatively new phenomena.

The solutions of singular perturbation problems involving differential equations often depend on several widely different lengths of time scales. Such problems can be divided into two broad classes: layer problems, treated using the *method of matched asymptotic expansions* (MMAE), and multiple-scale problems, treated by the *method of multiple scales* (MMS). Prandtl's boundary layer theory for the high-Reynolds flow of a viscous fluid over a solid body is an example of a boundary layer problem, and the semiclassical limit of quantum mechanics is an example of a multiple-scale problem. Hunter presents few examples where he illustrates some basic issues in perturbations theory with simple algebraic equations, which we recommend the reader should refer to his lectures. More details and examples on this section's topic can be found in the book by Awrejcewicz and Krysko [45] in Chap. 3.

3.13 Eigenvalue Problems

Spectral perturbation theory studies show the spectrum of an operator is perturbed when the operator is perturbed. In general, this question is a difficult one, and subtle phenomena may occur, especially in connection with the behavior of the continuous spectrum of the operators. Here, we consider the simplest case of the perturbation in an eigenvalue [11]. Let $\boldsymbol{H}$ be a Hilbert space with an inner product $\langle \cdot, \cdot \rangle$ and $A^{\varepsilon} : D(A^{\varepsilon}) \subset \boldsymbol{H} \to \boldsymbol{H}$ a linear operator in $\boldsymbol{H}$, with domain $D(A^{\varepsilon})$, depending smoothly on a real parameter ε. We assume that:

(a) A^{ε} is self-adjoin so that

$$\langle x, A^{\varepsilon} y \rangle = \langle A^{\varepsilon} x, y \rangle \text{for all } x, y \in D(A^{\varepsilon}).$$

(b) A^{ε} has a smooth branch of simple eigenvalues $\lambda^{\varepsilon} \in \mathbb{R}$ with eigenvectors $x^{\varepsilon} \in \boldsymbol{H}$, meaning that

$$A^{\varepsilon}x^{\varepsilon} = \lambda^{\varepsilon}x^{\varepsilon}. \tag{3.55}$$

We will compute the perturbation in the eigenvalue from its value at $\varepsilon = 0$ when ε is small but nonzero. A concrete example is the perturbation in the eigenvalues of a symmetric matrix. In that case, we have $\boldsymbol{H} = \mathbb{R}^n$ with the Euclidean inner product

$$\langle x, y \rangle = x^T y$$

in addition, $A^{\varepsilon} : \mathbb{R}^n \to \mathbb{R}^n$ is a linear transformation with and $n \times n$ symmetric matrix (a_{ij}^{ε}). The perturbation in the eigenvalues of a Hermitian matrix corresponds to $\boldsymbol{H}=\mathbb{C}^n$ with inner product $\langle x, y \rangle = \overline{x}^T y$. As we illustrate below with the Schrodinger equation of quantum mechanics, spectral problems for differential equations can be formulated in terms of unbounded operators acting in infinite-dimensional Hilbert spaces. We use the expansions

$$\begin{aligned} A^{\varepsilon} &= A_0 + \varepsilon A_1 + \cdots + \varepsilon^n A_n + \cdots \\ x^{\varepsilon} &= x_0 + \varepsilon x_1 + \cdots + \varepsilon^n x_n + \cdots \\ \lambda^{\varepsilon} &= \lambda_0 + \varepsilon \lambda_1 + \cdots + \varepsilon^n \lambda_n + \cdots \end{aligned}$$

in the eigenvalue problem of Eq. 3.55, equate coefficients of ε^n, and rearrange the result. We find that

$$(A_0 - \lambda_0 I)x_0 = 0, \tag{3.56}$$

$$(A_0 - \lambda_0 I)x_1 = -A_1 x_0 + \lambda_1 x_0, \tag{3.57}$$

$$(A_0 - \lambda_0 I)x_n = \sum_{i=1}^{n} \{-A_i x_{n-1} + \lambda_i x_{n-1}\}. \tag{3.58}$$

Assuming that $x_0 \neq 0$, Eq. 3.56 implies that λ_0 is an eigenvalue of A_0 and x_0 is an eigenvector. Equation 3.9 is then a singular equation for x_1. The following proposition gives a simple, but fundamental, solvability condition for this equation [11].

3.14 Quantum Mechanics

One application of this expansion is in quantum mechanics, where it can be used to compute the change in the energy levels of a system caused by a perturbation in its Hamiltonian.

The Schrodinger equation of quantum mechanics is

$$i\hbar\psi_t = H\psi.$$

Here t denotes time and $\hbar$ is Planck's constant. The wave function $\psi(t)$ takes values in a Hilbert space $\boldsymbol{H}$, and H is a self-adjoint linear operator acting in $\boldsymbol{H}$ with the dimensions of energy, called the Hamiltonian.

Energy eigenstates are wave functions of the form

$$\psi(t) = \mathrm{e}^{-iEt/h}\varphi,$$

where $\varphi \in \boldsymbol{H}$ and $E \in \mathbb{R}$. It follows from the Schrodinger equation that

$$H\varphi = E\varphi.$$

Hence, E is an eigenvalue of H and φ is an eigenvector. One of Schrodinger's motivations for introducing his equation was that eigenvalue problem led to the experimentally observed discrete energy levels of atoms.

Now suppose that the Hamiltonian

$$H^\varepsilon = H_0 + \varepsilon H_1 + O(\varepsilon^2)$$

depends smoothly on a parameter ε. Then, rewriting the previous result, we find that the corresponding simple energy eigenvalues (assuming they exist) have the expansion:

$$E^\varepsilon = E_0 + \varepsilon\frac{\langle\varphi_0, H_1\varphi_0\rangle}{\langle\varphi_0, \varphi_0\rangle} + O(\varepsilon^2),$$

where φ_0 is an eigenvector of H_0.

For example, the Schrödinger equation that describes a particle of mass m moving in $\mathbb{R}^d$ under the influence of a conservative force field with potential V: $\mathbb{R}^d \to \mathbb{R}$ is

$$i\hbar\psi_t = -\frac{\hbar^2}{2m}\Delta\psi + V\psi.$$

Here, the wave function $\psi(x, t)$ is a function of space variable $x \in \mathbb{R}^d$ and time $t \in \mathbb{R}$. At fixed time t, we have $\psi(\cdot, t) \in L^2(\mathbb{R}^d)$, where

$$L^2(\mathbb{R}^d) = \left\{u : \mathbb{R}^d \to \mathbb{C} \middle| u \text{ is measurable and } \int_{\mathbb{R}^d} |u|^2\, dx < \infty\right\}.$$

In the Hilbert space of square-integrable functions with inner product

$$\langle u, v\rangle = \int_{\mathbb{R}^d} \bar{u}(x)v(x)dx.$$

The Hamiltonian operator $H : \boldsymbol{D}(H) \subset \boldsymbol{H} \to \boldsymbol{H}$, with domain $\boldsymbol{D}(H)$, is given by

$$H = -\frac{\hbar^2}{2m}\Delta + V.$$

If u and v are smooth function that decay sufficiently rapidly at infinity, then Green's theorem implies that [11]

$$\begin{aligned}
\langle u, Hv\rangle &= \int_{\mathbb{R}^d} \bar{u}\left(-\frac{\hbar}{2m}\Delta v + Vv\right)dx \\
&= \int_{\mathbb{R}^d} \bar{u}\left\{\frac{\hbar^2}{2m}\nabla.(v\nabla\bar{u} - \bar{u}\nabla v) - \frac{\hbar^2}{2m}(\Delta\bar{u})v + V\bar{u}v\right\}dx \\
&= \int_{\mathbb{R}^d} \overline{\left(-\frac{\hbar^2}{2m}\Delta u + Vu\right)} v dx \\
&= \langle Hu, v\rangle.
\end{aligned}$$

Thus, this operator is formally self-adjoint. Under suitable conditions on the potential V, the operator can be shown to be self-adjoint with respect to an appropriately chosen domain [11].

Now suppose that the potential V^ε depends on a parameter ε and has the expansion

$$V^\varepsilon(x) = V_0(x) + \varepsilon V_1(x) + O(\varepsilon^2).$$

The perturbation in a simple energy eigenvalue

$$E^\varepsilon = E_0 + \varepsilon E_1 + O(\varepsilon^2)$$

assuming one exists is given by

$$E_1 = \frac{\int_{\mathbb{R}^d} V_1(x)|\varphi_0(x)|^2 dx}{\int_{\mathbb{R}^d} |\varphi_0(x)|^2 dx},$$

where $\varphi_0 \in L^2(\mathbb{R}^d)$ is an unperturbed energy eigenfunction that satisfies

$$-\frac{\hbar^2}{2m}\varphi'' + \frac{1}{2}kx^2\varphi = E\varphi.$$

Most of this section was borrowed from Hunter's reference [46].

Example 1 The one-dimensional simple harmonic oscillator has potential

$$V_0(x) = \frac{1}{2}kx^2$$

The eigenvalue problem

$$= \frac{\hbar^2}{2m}\varphi'' + \frac{1}{2}kx^2\varphi = E\varphi \quad \varphi \in L^2(\mathbb{R})$$

is exactly solvable. The energy eigenvalues are

$$E_n = \hbar\omega\left(n + \frac{1}{2}\right) n = 0, 1, 2, \cdots$$

where

$$\omega = \sqrt{\frac{k}{m}}$$

is the frequency of the corresponding classical oscillator. The eigenfunctions are

$$\varphi_n(x) = H_n(\alpha x)\mathrm{e}^{-\alpha^2 x^2/2},$$

where H_n is the Hermite polynomial,

$$H_n(\xi) = (-1)^n \mathrm{e}^{\xi^2} \frac{d^n}{d\xi^n} \mathrm{e}^{-\xi^2}$$

in addition, the constant α, with dimensions of 1/length, is given by

$$\alpha^2 = \frac{\sqrt{mk}}{\hbar}.$$

The energy levels E_n^ε of a slightly harmonic oscillator with potential

$$V^\varepsilon(x) = \frac{1}{2}kx^2 + \varepsilon\frac{k}{\alpha^2}W(\alpha x) + O(\varepsilon^2) \quad \text{as} \quad \varepsilon \to 0^+,$$

where $\varepsilon > 0$ have the asymptotic behavior

$$E_n^\varepsilon = \hbar\omega\left\{n + \frac{1}{2} + \varepsilon\Delta_n + O(\varepsilon^2)\right\} \quad \text{as} \quad \varepsilon \to 0^+,$$

where

$$\Delta_n = \frac{\int W(\xi) H_n^2(\xi) e^{-\xi^2} d\xi}{\int H_n^2(\xi) e^{-\xi^2} d\xi}.$$

For an extensive and rigorous discussion of spectral perturbation theory for linear operators, refer to references provided at the end of this chapter.

3.15 Summary

The solutions that are provided by the scientist and researcher in mathematics, physics, and science of engineering to the problem of instantaneous heat source, intense thermal wave and shock wave, or very strong and intense explosion and implosion throughout in this book and other references are important feature of self-similarity. As was mentioned in various sections, other mathematician or physicist were quoted "A time dependent phenomenon is called self-similar if the spatial distributions of the properties at different times can be obtained from another by a similarity transformation." According to Barenblatt [6], it was defined by a relationship below; if we define a time-dependent scale $r_0(t)$ for the spatial variable and $u_0(t)$ for any property u of the phenomenon (u can be a vector quantity, see Eq. 3.49), then the distribution of u at various instants can be expressed in the form

$$u = u_0(t) f\left(\frac{r}{r_0(t)}\right). \tag{3.59}$$

If we describe this distribution in self-similar coordinates $u/u_0(t)$ and $r/r_0(t)$, then the distributions for any value of time within the range considered is represented by a single curve, therefore in case of each example that was mentioned such as instantaneous heat source problem [38]:

$$r_0(t) = (\kappa t)^{1/2} \qquad u_0(t) = \theta_0(t) = Q/(\kappa t)^{1/2}. \tag{3.60}$$

For very intense thermal waves, we can show [38]

$$r_0(t) = \left[\left(\frac{E}{c}\right)^n (\kappa t)\right]^{1/(3n+2)} \qquad u_0(t)\left[\left(\frac{E}{c}\right)^2 (\kappa t)^{-3}\right]^{1/(3n+2)}. \tag{3.61}$$

Note that here $u_0(t)$ is the temperature scale, while in case of very intense blast (Taylor) [1], the problem is given in the following form:

$$
\begin{aligned}
r_0(t) &= \left(Et^2/\rho_0\right)^{1/5} & & \\
u_0(t) &= \rho_0 & & \text{(for density)} \\
u_0(t) &= (\rho_0)^{3/5}E^{2/5}t^{-6/5} & & \text{(for pressure)} \\
u_0(t) &= \left(Et^{-3}/\rho_0\right)^{1/5} & & \text{(for velocity)}.
\end{aligned} \tag{3.62}
$$

Finding a solution utilizing a self-similarity method for a mathematical and physics problem was a fascinating subject for these types of group of researchers and was considered as a very successful step. Self-similar method has many applications in the different field of science, and in particular dealing with nonlinear partial differential equations would allow reducing it to one that involves an ordinary differential equation for problems that encounter such PDEs during the time that numerical analysis and availability of computational platforms were not there.

Self-similar solutions provide some of the greatest simplifications to one-dimensional flows. Self-similarity allows the reduction of the partial differential equations, which contain two independent variables space and time, into a set of ordinary differential equations ODEs, where the single independent variable is a combination of space and time. The ODEs are then relatively easy to solve numerically or even analytically in some cases. They describe the asymptotic behavior of one-dimensional flows in a variety of circumstances, typically far away from the initial conditions and provided that the boundary conditions contain no spatial scale. Some exceptions apply, for example, self-similarity can prevail in exponential density gradient in planar geometry.

Today most of such nonlinear partial differential equations are solved using numerical method along with many software routines that are written around such problems, especially when one deals with Arbitrary Lagrangian/Eulerian boundary conditions. Moreover, self-similar solutions have been widely used as a method for evaluating all kinds of approximation techniques where few are mentioned throughout this chapter.

With well-defined self-similarity of first and second kind by Barenblatt and Zel'dovich [43], both the Taylor and Guderley problem were understood much better. In case of Taylor's (sometimes known as Sedov–Taylor) [1] solution, a strong explosion in which a strong shock wave propagates into the cold surroundings whose density profile decreases as $\rho \propto r^{-k}$ (power law), they used conservation of energy to obtain the scaling of the shock radius as a function of time. This type of approach or solutions are called first kind, while Guderley [5] found a self-similar solution of second kind describing imploding shock waves in a constant density environment both in spherical and cylindrical form. In contrast to the strong explosion problem, energy considerations cannot be used to deduce the scaling of the shock radius as a function of time. Instead, the scaling of the radius as a function of time must be found by demanding that the solution pass through a singular point of the equation. Such solutions are called self-similar solutions of the second kind.

Although the self-similar solutions not necessary provide an exact solution of isolated form for a specific problem, but above all as intermediate-asymptotic representations of the solutions of much wider classes of problem.

References

1. G.I. Taylor, The formation of a blast wave by a very intense explosion. I. Theoretical discussion. Proc. R. Soc. A **201**, 159–174 (1950)
2. P.L. Sachdev, *Shock Waves and Explosions* (Chapman & Hall/CRC, 2004).
3. G.I. Taylor, The formation of a blast wave by a very intense explosion. II. The atomic explosion of 1945. Proc. R. Soc. A **201**, 175–186 (1950)
4. Y.B. Zel'dovich, Y.P. Raizer, *Physics of Shock Waves and High-Temperature Hydrodynamic Phenomena*, vol 2 (Academic, New York, 1967)
5. W.D. Hayes, The propagation upward of the shock wave from a strong explosion in the atmosphere. J. Fluid Mech. **32**, 317 (1968)
6. D.D. Laumbach, R.F. Probestein, A point explosion in an old exponential atmosphere. J. Fluid Mech. **35**, 53 (1969)
7. L. Sedov, *Similarity and Dimensional Methods in Mechanics* (Academic, New York, 1969)
8. G. Guderley, Starke kugelige und zylindrische Verdichtungsstosse in der Nahe des Kugelmittelpunktes bzw. der Zylinderachse. Luftfahrtforschung **19**, 302–312 (1942)
9. S. Matsumora, O. Onodera, and I.L. Takayama, Noise induced by weak shock waves in automobile exhaust systems (Effects of viscosity and back pressure), *Proceedings of the 19th Int. Symp. on Shock Waves and Shock Tubes, Marseille, France*, Vol. III (1993), pp. 367–372
10. K.C. Phan, On the performance of blast deflectors and impulse attenuators, *Proceedings of the 18th Int. Symp. on Shock Waves and Shock Tubes, Sendai, Japan* (1991), pp. 927–934.
11. K.P. Stanyukovich, *Unsteady Motion of Continuous Media, Gastekhizdat, Englo. Transl.* (Pergamon Press, New York, 1969)
12. N.C. Freeman, On the stability of plane shock waves. J. Fluid Mech. **2**, 397–411 (1957)
13. K.C. Lapworth, An experimental investigation of the stability of plane shock waves. J. Fluid Mech. **6**, 469–480 (1959)
14. M.G. Briscoe, A.A. Kovitz, Experimental and theoretical study of the stability of plane shock waves reacted normally from perturbed hat walls. J. Fluid Mech. **31**, 529–546 (1968)
15. W.K. Van Moorhem, A.R. George, On the stability of plane shocks. J. Fluid Mech. **68**, 108 (1975)
16. M.I. Lighthill, Proc. R. Soc. A **198**, 454 (1949)
17. D.S. Butler, Converging Spherical and Cylindrical Shocks, *Armament Research Establishment*, Report No. 54/54 (1954)
18. A. Ramu, M.P. Ranga Rao, Converging spherical and cylindrical shock waves. J. Eng. Math. **27**, 411–417 (1993)
19. S.D. Ramsey, J.R. Kamm, J.H. Bolstad, The guderley problem revised. Int. J. Comput. Fluid Dyn. **26**(2), 79–99 (2012)
20. W. Chester, The propagation of shock waves in a channel of non-uniform width. Quart J. Mech. Appl. Math. **6**(4), 440–452 (1953)
21. R.F. Chisnell, The motion of a shock wave in a channel, with applications to cylindrical and spherical shock waves. J. Fluid Mech. **4**, 286–298 (1958)
22. G.B. Whitham, On the propagation of shock waves through regions of nonuniform area of flow. J. Fluid Mech. **4**(1), 337–360 (1958)
23. G.B. Whitham, A New approach to problems of shock dynamics. Part 1: Two-dimensional problems. J. Fluid Mech. **2**(1), 145–171 (1957)

24. G.B. Whitham, A New approach to problems of shock dynamics. Part II: Three-dimensional problems. J. Fluid Mech. **5**(1), 369–386 (1959)
25. K. Fong, B. Ahlborn, Stability of converging shock waves. Phys. Fluids **22**(3), 416–421 (1979)
26. J.H. Gardner, D.L. Book, I.B. Bernstein, Stability of imploding shocks in the CCW approximation. J. Fluid Mech. **1**(14), 41–58 (1982)
27. M. El-Mallah, *Experimental and Numerical Study of the Bleed Effect on the Propagation of Strong Plane and Converging Cylindrical Shock Waves*, The Department of Mechanical Engineering, Presented in partial fulfillment of the requirements for the degree of Doctor of Philosophy at Concordia University Montreal, Quebec, Canada, May 1997
28. P.L. Sachdev, *Self-Similarity and Beyond, Exact Solutions of Nonlinear Problems* (Chapman & Hall/CRC, 2000)
29. http://www.nuclear-knowledge.com/nuclear_physics.php
30. B.C. Reed, *The physics of the Manhattan project*, 3rd edn. (Springer, New York, 2015)
31. G. Guderley, Powerful spherical and cylindrical compression shocks in the neighborhood of the center of the sphere and of the cylindrical axis. Luftfahrtforschung **19**, 302 (1942)
32. J. Nuckolls, L. Wood, A. Thiessen, G. Zimmerman, Laser compression of matter to super-high densities: thermonuclear (CRT) applications. Nature **239**, 139 (1972)
33. L. Landau, E. Lifshitz, *Statistical Physics* (Addison-Wesley, Reading, 1969). Chap. 5
34. J. Mayer, M. Mayer, *Statistical Mechanics, 385* (Wiley, New York, 1940)
35. A.H. Shapiro, *The Dynamics and Thermodynamics of Compressible Fluid, Volume I and II*, 1st edn. (Ronald Press Company, New York, 1953)
36. A. Gsponer, *Fourth Generation Nuclear Weapons: Military Effectiveness and Collateral Effects* (Independent Scientific Research Institute, Geneva, Switzerland, 2008)
37. J. Becker, Testing the physics of nuclear isomers, Sci. Technol. Rev. (Lawrence Livermore National Laboratory, 2005) pp. 24–25
38. G.I. Barenblatt, *Scaling, Self-similarity, and Intermediate Asymptotics* (Cambridge University Press, Cambridge, 1996)
39. G. Birkhoff, *Hydrodynamics* (Princeton University Press, Princeton, NJ, 1950). Chap. V
40. L. Dresner, *Similarity Solutions of Nonlinear Partial Differential Equations* (Pitman Advanced Publishing Program, 1983)
41. K.T. Vu, J. Butcher, J. Carminati, Similarity solutions of partial differential equations using DESOLV. Comput. Phys. Commun. **176**(11–12), 682–693 (2007)
42. M. Van Dyke, *Perturbation Methods in Fluid Mechanics* (The Parabolic Press, Stanford, California, 1975)
43. G.I. Barenblatt, Y.B. Zel'dovich, *Self-Similar Solutions as Intermediate Asymptotics*, Institute of Mechanics, Moscow University, Moscow, USSR
44. G.I. Barenblatt, *Scaling*, 2nd edn. (Cambridge University Press, Cambridge, 2006)
45. J. Awrejcewicz and V.A. Krysko, *Introduction to Asymptotic Methods* (Chapman & Hall/CRC, Taylor & Francis Group, 2006)
46. J.K. Hunter, *Asymptotic Analysis and Singular Perturbation Theory* (University of California at Davis, Davis, 2004)
47. R. Serber, R. Rhodes, *The Los Alamos Primer*, vol 2 (University of California Press, Berkeley, 1992)

Chapter 4
Similarity Methods for Nonlinear Problems

In any mechanical studying either motion or heat transfer phenomena, many number of concepts are considered by introducing form of energy or defining velocity, stress, type of heat transfer, etc. in case of concerning motion and equilibrium, for example, can be formulated as problems for determining certain functions and numerical values for parameters that is characterizing such phenomena. Dealing with such problems and trying to solve them, we need to present certain rules and laws of mathematics and physics to relate certain nature of such event in a form of functional equations, which we know them as differential equations.

4.1 Similarity Solutions for Partial and Differential Equations

Despite numerous individual works on the subject, in particular similarity solutions of nonlinear partial differential equations (PDEs), the method of similarity solutions for solving nonlinear partial differential equations is still not as widely known as equally fruitful methods for solving linear partial differential equations using standard methods such as separation of variables or Laplace or Fourier transforms, for example. This section will aim to touch on the method of similarity solution by showing it as a practical technique and encourages the readers to study specific books on this topic for expanded knowledge and details on this subject [1–3].

It is of vital interest to say that solutions of certain sets of partial differential equations occurring in applied field such as exact solution of the boundary-layer equations of fluid mechanics can be found quite readily in spite of the failure of the more common classical approaches to yield any results. This will happen by employing transformations that reduce the system of partial differential equations to a system of ordinary differential equations. These types of solutions are generally

B. Zohuri, *Dimensional Analysis Beyond the Pi Theorem*,
DOI 10.1007/978-3-319-45726-0_4

designated as *similarity solutions* for reasons that are shown in following sections of this chapter as examples that are produced from different references and source that this author has found.

Nonlinear problem has always been an obstacle for scientists and engineers to deal with and be able to solve them in a closed form in particular in the early century where the computers were not available as their present forms to take numerical approaches in order solve by some means of perturbation or iteration analyses. So in most of the time, they end up eluding and trying to find an exact solution treatment. A great majority of nonlinear problems are described by systems of nonlinear partial differential equation together with appropriate boundary and initial conditions in order to model or scale some physical phenomena in particular in the field of fluid mechanics and gas dynamics as well as heat diffusion problems. In the early days of nonlinear science due to lack of computer platforms, attempts were made to reduce the system of PDEs to ordinary differential equations (ODEs) by the so-called similarity transformation. The ODEs could be solved by some closed method using techniques that were available to scientist. With the power of computer these days, the scenario has since then changed drastically. The nonlinear PDE systems with appropriate initial boundary conditions can now be solved effectively by means of sophisticated numerical methods and computers, with proper attention to the accuracy of solutions. The quest for exact solutions is now motivated by the desire to understand the mathematical structure of the solutions and, hence, a better understanding of the physical phenomena described by them. Analysis, computation, and not insignificantly intuition all pave the way to their discovery.

The similarity solutions in the earlier years were found by direct physical and dimensional arguments. The two very famous examples are point explosion (blast) and implosion, and converging shock to a focal point or axis of cylindrical geometry problems by Taylor (1950) [4, 5], Sedov (1959) [6], and Guderley (1942) [7] is among the examples of such approach.

Simple scaling discussion to induce similarity solutions was greatly presented by Zel'dovich and Raizer [8], and later on Barenblatt [9] amplified their work and clearly explained the nature of self-similar solutions of the first and second kind. This author was lucky enough to be able to participate in Barenblatt lecture, while it was given as an official course under the topic of dimensional analysis at the University of California at Berkeley during fall of 2008. Most importantly he manifested the role of these solutions as intermediate asymptotic, where the asymptotic behavior of solutions of these types of problem in a physical system of interest no longer depends on the details of the initial and boundary conditions for large time or distance for the system far from being in a limiting state. The early investigators relied greatly upon the physics of the problem to arrive at the similarity form of the solution and, hence, the solution itself.

Following Barenblatt and Zel'dovich [10], we call self-similar solutions that can be constructed using dimensional analysis alone as *self-similar solutions* of the *first or second kind*. There is complete similarity in all the parameters and variables, independent as well as dependent. As it was discussed, *self-similar solutions of the second kind* are self-similar with incomplete similarity in the dependent variables. Zel'dovich and Barenblatt 32 connect them to eigenvalue problems that are discussed extensively. If the problem contains no other length scale than the spatial variable x itself and no other time scale than the time variable t itself, dimensionless groups can only occur by combinations of x and t. As a result, the spatial distribution of the solution develops in time but remains geometrically self-similar [9, 11].

In this section we will show how in certain special cases the dimensional analysis can be used to obtain the same results that are deduced from group theory [1] by solving simple heat diffusion equation. This is an extension what years ago Boltzmann used the algebraic symmetry of the partial differential diffusion equation to study diffusion. This would be done with a concentration-dependent diffusion coefficient, although he never made any explicit mention of transformation groups or symmetry, but his approach and procedure were the same as the ones we know today. The bulk of his method was using the symmetry to find special solutions of the *partial* differential equation by solving a related *ordinary* differential equation.

Others that depended on the algebraic symmetry of the diffusion equation and could be generalized to other partial differential equations, including nonlinear ones, recognized Boltzmann's procedure later on. As matter of fact, Garrett Birkhoff [12] later on showed, using the algebraic symmetry of the partial differential equation, how solution can be found merely by solving a related ordinary differential equation, a much easier task to carry on. For reasons that become clear later, such solutions are now called similarity solutions.

Great attentions were paid by others after Birkhoff's [12] work to other fields of science using similarity solutions in such diverse fields as heat and mass transfer, fluid dynamics, solid mechanics, applied superconductivity, and plasma physics.

In most of these problems, two common features recurred. First, the transformation group leaving the partial differential equation invariant was *families* of *stretching* groups G of the form $x' = \lambda^{\alpha}x$, $y' = \lambda^{\beta}y$, and so on when $0 < \lambda < \infty$. Second, the partial differential equations were of second-order of Lie's second theorem. Note that G is the family of stretching groups to which the PDE is invariant and can be designated as *principle group*, and the second-order ODE obtained with the help of the principle group will be called the *principle* differential equation. The stretching group G to which the principle ODE is invariant will be called the *associated group*, and the first-order ODE obtained with the help of the associated group will be called the *associated* differential equation. Figure 4.1 summarizes this scheme of terminology [2].

Second feature-type problems suggest that the related ordinary differential equation is also of second order. In some cases, this second-order ODE is integrable in terms of elementary or tabulated functions, but in most cases it is not. It also was observed by Dresner [13] that in problems with the first feature also, the second-order ODE is invariant to a stretching group G related to the family of using Lie's second theorem mentioned above; then the second-order ordinary differential

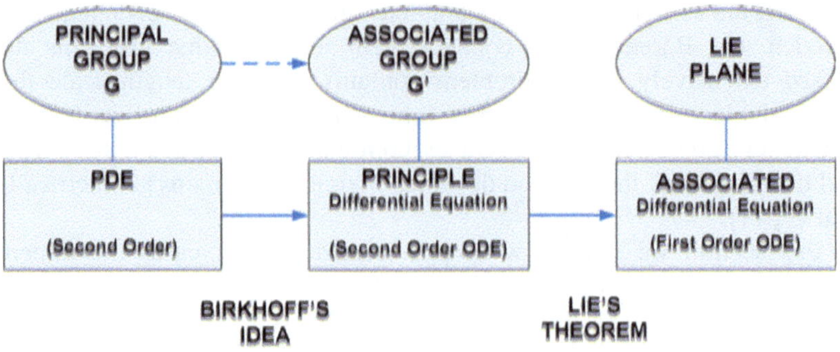

Fig. 4.1 Scheme of terminology

equation can be reduced to first order by changing of variables. The first-order equation can be analyzed very conveniently by studying its direction field [2]. For further information, one should refer to Dresner [2] book.

4.2 Fundamental Solutions of the Diffusion Equation Using Similarity Method

We know fundamentally by solving the heat equation, one gets the temperature distribution in an infinite medium due to the concentrated addition of a source heat. For simplicity, first we take under consideration one-dimensional heat conduction in an infinite bar with constant properties and produce a governing diffusion equation along with its boundary and initial conditions. Initially at $t = 0$, the bar is at uniform temperature $T(x, 0) = T_0 = 0$. The governing equation in a mathematical form along with its boundaries is presented as follows:

$$\rho c \frac{\partial T(x,t)}{\partial t} - k\frac{\partial T^2(x,t)}{\partial x^2} = Q\delta(x)\delta(t), \tag{4.1}$$

where boundary and initial conditions are

$$\begin{aligned} &-\infty < x < \infty \\ &t \geq 0 \\ &T(x,0) = 0 \\ &T(\pm\infty, t) \to 0. \end{aligned}$$

In Eq. 4.1 we have assume the source is the origin merely x_0, but if we chose any other location such as x along the boundary, then we need to replace our coordinate by $x - x_0$, and it is evident that it follows formally from the translation invariance. If the temperature is restricted in a physically reasonable way (e.g., bounded below for $Q > 0$), then the problem specified by Eq. 4.1 has a unique solution. Considering above heat diffusion setup, the fundamental principle of dimensional analysis enables us a form of solution for this case. Based on this fundamental principle and dimensional analysis role, we can state that

Every problem must be able to be expressed in form of dimensionless variables or alternatively all equalities must involve only dimensionally consistent quantities.

The fundamental principle is a statement of invariance of all physical problems with respect to choice of units of measurement. The quantities entering above problem and Eq. 4.1 have physical dimensions as follows, and we use the associate units per our established rules in Chap. 1:

Physical constant	*Units*
ρ = density	$\{\rho\} = \frac{\text{Kg}}{M^3}$
c = specific heat	$\{c\} = \frac{\text{cal}}{\text{deg}\cdot K}$
k = thermal conductivity	$\{k\} = \frac{\text{cal}}{SM^2}\cdot\frac{M}{\text{deg}} = \frac{\text{cal}}{\text{deg}SM}$
Q = heat added per cross-section area	$\{Q\} = \frac{\text{cal}}{M^2}$
Variables	
T = temperature	$\{T\} = \text{deg}$
x = space coordinate	$\{x\} = M$
t = time coordinate	$\{t\} = S$

Per our general principle definition, the independent variables (x, t) should be expressed in dimensionless variables, but an examination of the physical constants shows that no combination of them can provide a physical constant with dimensions either of length (M) or time (S) which could be used to make x or t dimensionless.

Note that if a characteristic temperature T_c exists, then (x, t) could be made dimensionless. We are implicitly using the fact that the initial temperature T is not characteristic and only $(T - T_0)$ can appear [1].

The problem has no characteristic length of time scale. Note, however, the thermal diffusivity κ:

$$\kappa = \frac{k}{\rho c} \{k\} = \frac{M^2}{S}.$$

Hence a dimensionless variable can only be formed from a suitable combination of (x, t), namely,

$$z = \frac{x}{2\sqrt{\kappa t}}. \tag{4.2}$$

In order to make the dependent variable T dimensionless, a quantity with the dimensions of degree must be formed with the help of the physical constants. The quantity, $Q/\rho c$, has dimensions

$$\left\{\frac{Q}{\rho c}\right\} = M \cdot \text{deg}.$$

Since no characteristic temperature exists, no quantity with dimensions of purely (deg) can be found. However, by using the length x or $\sqrt{\kappa t}$, a dimensionless combination can be formed as the following equation:

$$\frac{T\sqrt{\kappa t}}{Q/\rho c}. \tag{4.3}$$

Thus, the functional form of the solution is completely defined as Eq. 4.4:

$$T(x, t, \rho c, k, Q) = \frac{Q}{\rho c\sqrt{\kappa t}} f(z), \tag{4.4}$$

where $f(z)$ is a dimensionless function. $f(z)$ must satisfy an ordinary differential equation [1].

The analogous problem in two or three dimensions is different only in the dimensions of the heat addition term. Thus, two- and three-dimension cases can be described as the following equations:

$$\text{Two-dimension}\{Q\} = \frac{\text{heat added}}{\text{length}} = \frac{\text{cal}}{M}; T = \frac{Q}{\rho c\kappa t} f(z_2) \tag{4.5}$$

$$\text{Three-dimension}\{Q\} = \text{heat added} = \text{cal}; T = \frac{Q}{\rho c(\kappa t)^{3/2}} f(z_3). \tag{4.6}$$

Both parameters z_2 and z_3 for cylindrical and spherical coordinates are defined per following relationships:

$$z_2 = \frac{\sqrt{x^2 + y^2}}{2\sqrt{\kappa t}} = \frac{r}{2\sqrt{\kappa t}} \quad r = \text{cylindrical radius}$$

$$z_3 = \frac{\sqrt{x^2 + y^2 + z^2}}{2\sqrt{\kappa t}} = \frac{r}{2\sqrt{\kappa t}} \quad r = \text{spherical radius}.$$

The connection between dimensional analysis and invariance of partial differential equations under stretching transformations is fully discussed by Bluman and Cole [1].

4.3 Similarity Method and Fundamental Solutions of the Fourier Equation

Since terms of a properly formulated equation in physics—a quantity equation—are dimensionally homogeneous, it must be possible to cast it in a dimensionless form. Therefore, if we consider a simple case of a linear ordinary and homogeneous equation as well as both its boundary and initial conditions, one can find a solution using dimensional analysis and similarity methods. This method allows us to identify the dimensionless arguments of the solution rather than seeking an explicit solution for such set of differential equation. To further enhance our understanding about such method, we look at Eq. 4.1 and consider a simple one-dimensional case of it.

For simplicity we look at diffusion equation that is known as *Fourier equation,* and in one-dimensional form, it can be written as

$$\frac{\partial T(x,t)}{\partial t} = \kappa \frac{\partial T^2(x,t)}{\partial x^2}, \tag{4.7}$$

where κ is thermal diffusivity and is defined as in preceding section. Equation 4.7 is a homogeneous and linear partial equation of second order with constant coefficient (i.e., $\kappa = k/\rho c$ is constant). Dimensional consideration allows us to predict a solution that may exist in the form of

$$T(x,t) = \Theta\left(\frac{x}{2(\kappa t)^{1/2}}\right) = \Theta(\xi). \tag{4.8}$$

Considering such solution implies that temperature function T does not separately depend on both variable x and t, but on the composite variable $\xi = x/2(\kappa t)^{1/2}$. Substitution of such solution as Eq. 4.8 into Eq. 4.7 reduces the partial differential equation to an ordinary differential equation as below:

$$\Theta'' + 2\xi\Theta' = 0. \tag{4.9}$$

Equation 4.9 may possess a general solution of the following form:

$$\Theta = C_1 \int_0^{x/2\left((\kappa t)^{1/2}\right.} \exp\left(-\xi^2\right) + C_2. \tag{4.10}$$

Now if we define Gauss's error integral as follows:

$$\mathrm{erf}(z) = \frac{2}{\pi^{1/2}} \int_0^2 \exp(-\xi^2) d\xi, \tag{4.11}$$

which in short we called error function, we bear the following properties as well:

$$\begin{cases} \mathrm{erf}(0) = 0 \\ \mathrm{erf}(\infty) = 1 \\ \mathrm{erf}(-z) = -\mathrm{erf}(z) \end{cases}.$$

If we conveniently choose the constants C_1 and C_2, we obtain the special solution:

$$T(x,t) = \frac{2T_c}{\pi^{1/2}} \int_0^{x/2(\kappa t)^{1/2}} \exp(-\xi^2) d\xi = T_c \mathrm{erf}\left(\frac{x}{2(\kappa t)^{1/2}}\right). \tag{4.12}$$

The preceding solution of Eq. 4.7 satisfies the following boundary and initial conditions:

$$\begin{aligned} T(x,t) &= T_c && \text{for all } x > 0 \text{ and } t = 0. \\ T(x,t) &= -T_c && \text{for all } x < 0 \text{ and } t = 0. \\ T(x,t) &= 0 && \text{for all } t > 0 \text{ and } x = 0. \end{aligned}$$

Since our Eq. 4.7 is linear, therefore, any derivatives of solution in respect to x and t are also solutions. Now if we differentiate Eq. 4.10 with respect to x, we get the following:

$$T = \frac{C}{2(\kappa t)^{1/2}} \exp(-\xi^2), \tag{4.13}$$

and if we keep on repeating the differentiation again, we obtain the following:

$$T = -\frac{Cx}{4(\kappa t)^{3/2}} \exp(-\xi^2). \tag{4.14}$$

Equations 4.13 and 4.14 are called fundamental solutions of Fourier Eq. 4.7, and their application may be considered in a thermal explosion if we consider a point source. In this case, a sudden source of heat is released in a volume whose extent can be neglected compared with the dimensions of surroundings of interest, such as nuclear explosion. This application has been discussed in Appendix E along with more general form of Fourier equation that is introduced.

As far as differential equations such as Fourier Eq. 4.7 as well as their boundary and initial conditions are concerned, the preceding steps apply to the variables as well as the coefficients. By virtue of this procedure, we can identify the dimensionless arguments of the solution without actually finding an explicit solution itself. Most often the number of dimensionless arguments is smaller than the number of dimensional quantities, which we encounter in problems like that.

Now take Eq. 4.7 and apply the above statements and assume $\Theta = T/T_c$ and $\xi = x/X$; then we can define a dimensionless temperature Θ and a dimensionless length variable ξ, where T_c is a characteristic constant temperature and X a characteristic length. We need to choose these two elements that in a way that both boundary and initial conditions also become dimensionless. For instance, if we choose T_c from boundary conditions of our choice, the latter become a pure number—unity in this case. Similarly, it is useful to make X equal to half thickness of slap geometry. When we are dealing with a situation that no characteristic of time exists, then it is useful to introduce a dimensionless time in the form of $\tau = \kappa t/X^2$. A good example of such case is stepwise increase or decrease of temperature at the boundary, for instance, if this boundary is defined in the form of a time-dependent temperature $T_0(t)$ or if a time varying heat flux $\dot{q}\,(t)$ is considered in the problem. Under these circumstances it may not be possible to specify a single such dimensionless characteristic. With all above definitions and constraints, the Fourier Eq. 4.7 will obtain the dimensionless form of the following:

$$\frac{\partial \Theta}{\partial \tau} = \frac{\partial^2 \Theta}{\partial \xi^2} \quad \text{or} \quad \frac{T}{T_c} = f\left(\frac{x}{X}, \frac{\kappa t}{X^2}\right). \tag{4.15a}$$

The solution to this equation follows a homogeneous product type; if the problem and its geometry contain no characteristic length, X, then we can write

$$\left(\frac{x}{X}\right)^{\kappa} \quad \frac{\kappa t}{X^2}\,(\kappa \text{ is a constant}), \tag{4.15b}$$

which is an indication of being independent of X. This also can be verified if the proper relationship is in the form of

$$\left(\frac{\kappa t}{x^2}\right)^{\gamma}\,(\gamma \text{ is a constant}). \tag{4.15c}$$

In this case x and X have replaced each other. Under these conditions the solution we deduce will be in the following form:

$$\frac{T}{T_c} = f\left(\frac{\kappa t}{x^2}\right) \tag{4.16a}$$

or equivalently

$$\begin{cases} \dfrac{T}{T_c} = f\left(\dfrac{x}{2(\kappa t)^{1/2}}\right) \\ T = T_c f\left(\dfrac{x}{2(\kappa t)^{1/2}}\right) \end{cases}. \tag{4.16b}$$

As you can see, this exactly has the same result as we found in Eq. 4.11.

When we are dealing with problems (i.e., in heat exchanger design hot fluid versus cold one) that contain a boundary condition of the third kind (see Appendix F) with respect to a wall perpendicular to x, with the positive direction of x pointing inward, under assumption that the temperature of the fluid is $T(x,t)\big|_{x\to\infty} = T_\infty = 0$, then we would have

$$= \left(\frac{\partial T(x,t)}{\partial x}\right) = -\frac{T_{\text{wall}}}{k/h}. \tag{4.17}$$

Utilizing similar dimensionless variables as before, we can transform Eq. 4.17 to the form of

$$\frac{1}{\Theta_{\text{wall}}}\left(\frac{\partial \Theta}{\partial \xi}\right)_{\text{wall}} = \frac{hX}{k}. \tag{4.18}$$

If we apply the Buckingham's Pi theorem to heat conduction problems with appropriate boundary conditions, we will see that the original seven physical quantities reduce to four when introduction of dimensionless quantities takes place, simply because the problem of interest is characterized by four fundamental dimensions as length (L), time (t), temperature (T), and heat (Q). Note that one of the well-known dimensions, mass (M), plays no part in this problem.

We can reduce Eq. 4.16b to the following form and use the dimensionless numbers that are introduced earlier in previous chapter, and they are named after their contributors that discovered them and they are

$$\text{Fourier number} : \text{Fo} = \frac{\kappa t}{X^2} \quad \text{or} \quad \frac{\kappa t}{x^2}$$

$$\text{Biot number} : \text{Bi} = \frac{hX}{k} \quad \text{or} \quad \frac{hx}{k}.$$

Using the above notation, the new form of Eq. 4.16b is as follows:

$$\frac{T}{T_c} = f\left(\frac{x}{X}, \text{Fo}, \text{Bi}\right). \tag{4.19}$$

We will illustrate the concept of similarity solutions using the following examples:

Fig. 4.2 A semi-infinite bar heated from $x = 0$

Example 1: (The Heat Bar) Consider the heat conduction problem governed by the following one-dimensional conduction equation where T is temperature while x and t represent spatial coordinate and time, respectively. α is the diffusion coefficient of heated bar and it is the function of its material properties and we assume constant for this example:

$$\frac{\partial T}{\partial t} = \alpha \frac{\partial^2 T}{\partial x^2} 0 \leq x < \infty \quad \text{for} \quad t > 0.$$

In this example we assume a very long thermally isolated bar, initially at a uniform temperature zero, is heated at one end by a constant flux. There is no source at the other end; as a result the boundary and initial conditions associated with this problem and satisfying the above heat conduction are given by

$$\begin{aligned} &\frac{\partial}{\partial x} T(0,t) = -Q_0 && t > 0 \\ &T(x,0) = 0 && 0 \leq x < \infty \\ &0 \leq T(x,t) < \infty && x \to \infty \quad t > 0. \end{aligned}$$

The bar is modeled as semi-infinite (see Fig. 4.2), with a crosswise constant temperature distribution describe by above equation and related boundaries.

If we try to scale, we find that we have modeled any explicit length, time, or temperature scale out of our problem. Therefore, we can only make the problem dimensionless on the available implicit scales:

- As there is no length scale in x or t, the intrinsic scale can only be $\sqrt{\alpha t}$.
- The only temperature in the problem is $Q_0 x$ or $Q_0 \sqrt{\alpha t}$.

Therefore, we assume

$$T(x,t) = Q_0 x g(\eta),$$

where the similarity variable η is given by

$$\eta = \frac{x}{\sqrt{4\alpha t}}.$$

It follows that g satisfies the reduced ordinary differential equation as follows:

$$\frac{1}{2}\eta g'' + \left(1 + \eta^2\right) g' = 0.$$

With boundary conditions

$$\lim_{\eta\to\infty} g(\eta) = 0 \lim_{\eta\to 0} (g(\eta) + \eta g'(\eta)) = -1.$$

This has the solution of

$$g(\eta) = \frac{1}{\eta\sqrt{\pi}}\exp(-\eta^2) - \operatorname{erfc}(\eta),$$

where $\operatorname{erfc}(\eta) = \frac{2}{\sqrt{\pi}}\int_\eta^\infty e^{-\xi^2}\,d\xi$ is the complementary error function. Hence we have

$$T(x,t) = Q_0\left[\sqrt{\frac{4\alpha t}{\pi}}\exp(-\eta^2) - x\operatorname{erfc}(\eta)\right].$$

Note: There exists no stationary solution. The found solution is completely similar, both in the independent and in the dependent variables. Therefore it is a similarity solution of the first kind.

Example 2: (Convection) In the convection problem

$$\begin{aligned}&\frac{\partial u}{\partial t} + U_0\frac{\partial u}{\partial x} = 0 \quad x\in\Re \quad t>0\\ &u(x,0) = H(x) \qquad x\in\Re,\end{aligned}$$

there is the length scale given by $U_0 t$ and a length scale, say L, hidden in the initial profile $H(x)$, as x cannot occur on its own. The dimensions of u and H are the same, say H_0, and we write $H(x) = H_0 h\left(\frac{x}{L}\right)$. We scale $x = L\xi$, $t = \frac{L}{U_0}\tau$, $u = H_0 v$, and $v(\xi,0) = h(\xi)$ to get

$$\frac{\partial v}{\partial \tau} + \frac{\partial v}{\partial \xi} = 0$$

with solution $v(\xi,\tau) = h(\xi - \tau)$.

Further examples of this kind could be found in the book written by Mattheij et al. [14].

4.4 Fundamental Solutions of the Diffusion Equation: Global Affinity

We will extend our previous discussion by studying its invariance under global transformations, in particular stretching transformations which are presented by Bluman and Cole [1] which we have exactly here repeated. The method is thus analogous to the approach that is shown in their book under Sect. 1.1 in introducing the ideas for ordinary differential equations. In effect we see from their approach what can be done without the use of infinitesimal transformation.

Therefore, our Equation Set 4.1 can be written in the following if we let the temperature field be in form of

$$T(x,t) = \Theta(x,t) \tag{4.20}$$

so that

$$\begin{aligned} &\frac{\partial \Theta(x,t)}{\partial t} - \kappa \frac{\partial^2 \Theta(x,t)}{\partial x^2} = \frac{Q}{\rho c}\delta(x)\delta(t) \\ &\Theta(x,0) = 0 \\ &\Theta(\pm\infty, t) \to 0. \end{aligned} \tag{4.21}$$

In this analysis, we assumed that $T(x,t) = \Theta(x,t)$ is a solution for Eq. 4.1. In addition, we consider the general stretching transformation of the (T, x, t) space as follows:

$$\begin{aligned} T^* &= \tau T \\ x^* &= \alpha x \\ t^* &= \beta t \end{aligned} \tag{4.22}$$

with parameters (γ, α, β). If $\gamma(\beta)$, $\alpha(\beta)$ are somehow determined, Eq. 4.22 is a one-dimensional group of transformations with the identity element $\gamma = \alpha = \beta = 1$ (i.e., if $\beta = \mathrm{e}^{\varepsilon}$, then $\varepsilon = 0$ corresponds to the identity).

$$T^* = \Theta^*\left(x^*, t^*\right). \tag{4.23}$$

Now we ask how the original solution surface transforms [1]:

$$\frac{\partial \Theta(x,t)}{\partial t} = \frac{\beta}{\gamma}\frac{\partial \Theta^*\left(x^*,t^*\right)}{\partial t}, \ \frac{\partial \Theta(x,t)}{\partial x} = \frac{\alpha}{\gamma}\frac{\partial \Theta^*\left(x^*,t^*\right)}{\partial x^*}, \ \frac{\partial^2 \Theta(x,t)}{\partial x^2} = \frac{\alpha^2}{\gamma}\frac{\partial^2 \Theta^*\left(x^*,t^*\right)}{\partial x^{*2}}.$$

Since $\Theta(x,t)$ is defined by the set of Eq. 4.8, then we have [1]

$$\frac{\beta}{\gamma}\frac{\partial\Theta^*(x^*,t^*)}{\partial t^*}-\kappa\frac{\alpha^2}{\gamma}\frac{\partial^2\Theta^*(x^*,t^*)}{\partial x^{*2}}=\frac{Q}{\rho c}\delta\left(\frac{x^*}{\alpha}\right)\delta\left(\frac{t^*}{\beta}\right). \tag{4.24}$$

For invariance, it is necessary that both the operators on the left and the right-hand side of Eq. 4.24 agree with those in Eq. 4.8, multiplied by a common factor. Thus, for invariance [1]

$$\alpha^2=\beta\alpha=\sqrt{\beta}. \tag{4.25}$$

Further, it follows from the integral definition of the γ-functions [1]:

$$\int\delta(x)dx=1,\int\delta(ax)dx=\frac{1}{a}\int\delta(ax)d(ax)$$

that

$$\delta(ax)=\frac{1}{a}\delta(x). \tag{4.26}$$

Thus, Eq. 4.11 becomes

$$\frac{\beta}{\gamma}\left[\frac{\partial\Theta^*(x^*,t^*)}{\partial t^*}-\kappa\frac{\alpha^2}{\gamma}\frac{\partial^2\Theta^*(x^*,t^*)}{\partial x^{*2}}\right]=\beta^{3/2}\frac{Q}{\rho c}\delta(x^*)\delta(t^*). \tag{4.27}$$

For invariance of the equation, then

$$\gamma=\frac{1}{\sqrt{\beta}}. \tag{4.28}$$

Finally Bluman and Cole 5 argue that one-parameter family of transformations of the (T,x,t) space to itself is thus Eq. 4.9:

$$\begin{aligned}T^*&=\frac{1}{\sqrt{\beta}}T\\x^*&=\sqrt{\beta}x\\t^*&=\beta t.\end{aligned} \tag{4.29}$$

Equation 4.17 is the group of transformations leaving invariant problem of Eq. 4.1. Under this transformation, the initial and boundary conditions attached to Eq. 4.1 are also invariant. Thus for $\Theta^*(x^*,t^*)$ we have

$$\begin{aligned}
&\frac{\partial \Theta^*(x^*,t^*)}{\partial t^*} - \kappa \frac{\partial^2}{\gamma} \frac{\partial^2 \Theta^*(x^*,t^*)}{\partial x^{*3}} = \frac{Q}{\rho c}\delta(x^*)\delta(t^*) \\
&\Theta^*(x^*,t) = 0 \\
&\Theta^*(\pm\infty,t^*) \to 0.
\end{aligned} \tag{4.30}$$

Now, due to the uniqueness, $\Theta(x,t)$ must be the *same* function of (x^*,t^*) as $\Theta^*(x^*,t^*)$ is of (x,t), that is [1],

$$\Theta^*(x,t) = \Theta(x^*,t^*). \tag{4.31}$$

Because of the transformation Eq. 4.29 and the invariance condition Eq. 4.31, we thus obtain a functional equation, which must be satisfied by the solution [1]:

$$T^*(x^*,t^*) = \Theta(x^*,t^*) = \frac{1}{\sqrt{\beta}}T = \frac{1}{\sqrt{\beta}}T(x,t)$$

or

$$\Theta\left(\sqrt{\beta}x,\beta t\right) = \frac{1}{\sqrt{\beta}}\Theta(x,t). \tag{4.32}$$

Equation 4.32 holds for all values of β. From this functional relation, the functional form that the solutions are $\Theta(x,t)$ must have and can be deduced in various ways. For example, we can say that the factor $1/\sqrt{\beta}$ is a scaling like $1/\sqrt{t}$ from Eq. 4.29, and a coordinate like $x/\sqrt{t}$ is invariant so that necessarily [1]

$$\Theta(x,t) = \frac{1}{\sqrt{t}}f\left(\frac{x}{\sqrt{t}}\right). \tag{4.33}$$

Note that alternatively $\frac{x}{\sqrt{t}}$ and $\sqrt{t}\Theta$ are two functionally independent invariants of Eq. 4.29. The solution form of Eq. 4.33 is obtained by setting one invariant as an arbitrary function of the other [1].

Evidently, Eq. 4.32 is satisfied and within trivial changes of Eq. 4.21 is unique. Alternatively consider $\partial/\partial\beta$ of Eq. 4.32 near the identity ($\beta = 1$) (i.e., study the infinitesimal form of Eq. 4.32).

$$\frac{1}{2\sqrt{\beta}}x\frac{\partial \Theta}{\partial x}\left(\sqrt{\beta}x,\beta t\right) + t\frac{\partial \Theta}{\partial t}\left(\sqrt{\beta}x,\beta t\right) = \frac{1}{2\beta^{3/2}}\Theta(x,t). \tag{4.34}$$

As $\beta \to 1$, we see that $\Theta(x,t)$ must also satisfy a first-order partial differential equation:

$$\frac{1}{2}x\frac{\partial \Theta}{\partial x}(x,t) + t\frac{\partial \Theta}{\partial t} = -\frac{1}{2}\Theta(x,t). \tag{4.35}$$

This enables the form of the solution to be found since the general solution of Eq. 4.35 involves an arbitrary function. The characteristic equations associated with Eq. 4.35 are [1]

$$\frac{dx}{\frac{1}{2}x} = \frac{dt}{t} = \frac{d\Theta}{\frac{1}{2}\Theta}. \tag{4.36}$$

The integral of the first two is

$$\zeta = \frac{x}{2\sqrt{t}}. \tag{4.37}$$

In addition, the integral of the second two is

$$\begin{aligned} \log\Theta &= -\frac{1}{2}\log t + \log F(\zeta) \\ \Theta(x,t) &= \frac{1}{\sqrt{t}}F(\zeta), \quad \zeta = \frac{x}{2\sqrt{t}}. \end{aligned} \tag{4.38}$$

This agrees exactly with form Eq. 4.4 derived by dimensional analysis; introducing the dimensional quantities, let

$$T = \Theta(x,t) = \frac{Q}{\rho c\sqrt{\kappa t}}f(z), \quad z = \frac{x}{2\sqrt{\kappa t}}. \tag{4.39}$$

We now proceed to the solution by deriving the ordinary differential equation for $f(z)$. We have

$$\begin{aligned} \frac{\partial T}{\partial t} &= \frac{Q}{\rho c\sqrt{\kappa}}\frac{1}{}\left[\frac{1}{\sqrt{t}}\frac{df(z)}{dz}\left(-\frac{z}{2t}\right) - \frac{1}{2t^{3/2}f(z)}\right] \\ &= -\frac{Q}{\rho c\sqrt{\kappa}2t^{3/2}}\frac{1}{}\left[z\frac{df(z)}{dz} + f(z)\right] \\ \frac{\partial^2 T}{\partial x^2} &= \frac{Q}{\rho c4(\kappa t)^{3/2}}\frac{d^2 f(z)}{dz^2}. \end{aligned}$$

so that for $t \geq 0$, the partial differential equation of Eq. 4.8 becomes

$$\frac{d^2 f(z)}{dz^2} + 2z\frac{df(z)}{dz} + 2f(z) = 0. \tag{4.40}$$

Fig. 4.3 Presentation of conditions of Eq. 4.29

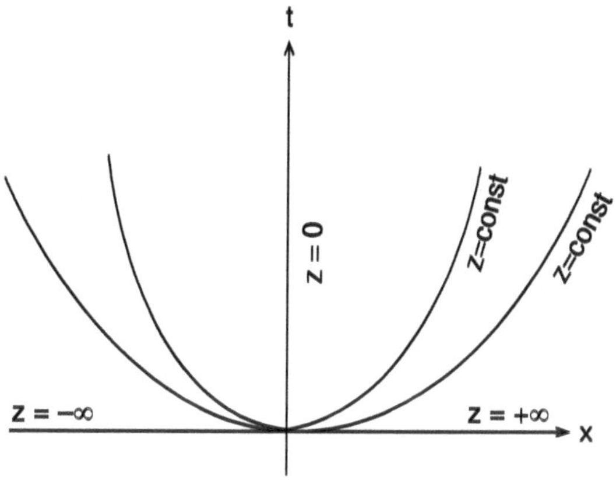

The boundary condition (BC) and initial condition (IC) of Eq. 4.8 are now presented as Eq. 4.41 below. See also Fig. 4.3.

$$f(z) \to 0 \text{ for } z = \pm\infty. \tag{4.41}$$

The condition in Eq. 4.41 is not sufficient to define the solution; therefore, account must be taken of the fact that an amount of heat Q has been introduced to the medium. The law of conservation of total heat or $\int_0^t dt \int_{-\infty}^{+\infty} dx$ of the heat conduction is

$$\rho c \int_{-\infty}^{+\infty} \Theta(x,t)dx = Q. \tag{4.42}$$

Using the similarity form of Eq. 4.39, then Eq. 4.42 becomes

$$\int_{-\infty}^{+\infty} \frac{1}{\sqrt{\kappa t}} f(z)dx = 1$$

or

$$\int_{-\infty}^{+\infty} f(z)dx = \frac{1}{2}. \tag{4.43}$$

Equation 4.43 is extra condition needed to define the solution uniquely. Now the solution to Eq. 4.28 can be expressed in terms of Hermite functions, and of course the solution to the problem is well known.

There are more details on the subject of Similarity Methods for Differential Equation and Dimensional Analysis given by Bluman and Cole [1] as well as other books and they are beyond the scope of this book to reflect them all, so we recommend the readers to refer to them.

Hermite Function or Polynomial Definition

In the **Sturm–Liouville boundary value problem**, there is a special case called Hermite's differential equation which arises in the treatment of the harmonic oscillator in quantum mechanics. **Hermite's differential equation** is defined as

$$\frac{d^2y}{dx^2} - 2x\frac{dy}{dx} + 2ny = 0$$

where n is a real number. For n is a nonnegative integer, i.e., $n = 0, 1, 2, 3, \cdots$, the solutions of Hermite's differential equation are often referred to as **Hermite polynomials** $H_n(x)$.

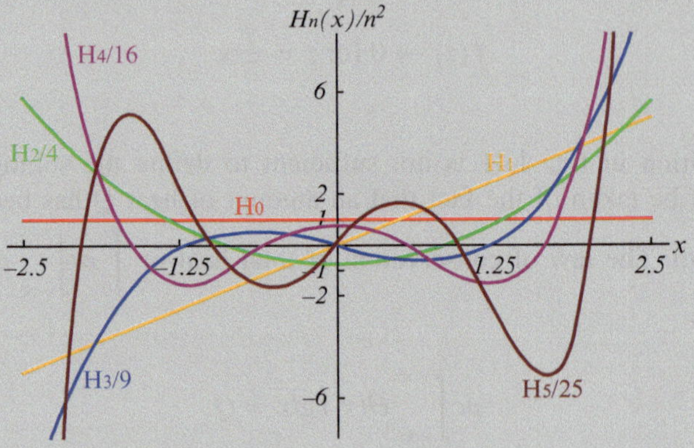

Plots of Hermite Polynomials

In general a family of orthogonal polynomials which arise as solutions to Hermite's differential equation, a particular case of the hypergeometric differential equation. In a special case of system of polynomials of successively increasing degree for $n = 0, 1, 2, 3, \cdots$, the Hermite polynomials $H_n(x)$ are defined by the formula

$$H_n(x) = (-1)^n e^{x^2} \frac{d^n}{dx^n} d^{-x^2}$$

In particular, $H_0 = 1$, $H_1 = 2x$, $H_2 = 4x^2 - 2$, $H_3 = 8x^3 - 12x$, and $H_4 = 16x^4 - 48x^2 + 12$.

(continued)

(continued)

In mathematics, the Hermite polynomials are a classical orthogonal polynomial sequence that arises in probability, such as the Edgeworth series; in combinatorics, as an example of an Appell sequence, obeying the umbral calculus; in numerical analysis as Gaussian quadrature; and in physics, where they give rise to the eigenstates of the quantum harmonic oscillator. They are also used in systems theory in connection with nonlinear operations on Gaussian noise. They are named in honor of Charles Hermite (1864) although they are actually due to Chebyshev (1859).

4.5 Solution of the Boundary-Layer Equations for Flow over a Flat Plate

A classical example of a similarity solution of a system of partial differential equations is the solution for the boundary-layer flow over a flat plate in a uniform system. This example was copied from Hansen book [15]. As part of a complete solution of a specific problem, we show a form of diffusion equation in previous section in term of the viscous flow momentum equation for fluid motion over a suddenly accelerated plane. An infinite plate is assumed to be immersed in an incompressible fluid, which is at rest. At time $t = 0$, the plate suddenly accelerated in its own plane to a constant velocity. The problem is to find the velocity of the fluid surrounding the plate as a function of time and distance from the plate (Fig. 4.4).

The application of Newton's second law of motion to a fluid particle leads to the following equation:

$$\frac{\partial u}{\partial t} = \nu \frac{\partial^2 u}{\partial y^2}, \tag{4.44}$$

where

$u =$ fluid velocity in x-direction.
$\frac{\partial u}{\partial x} =$ fluid acceleration.
$\nu \frac{\partial^2 u}{\partial y^2} =$ surface viscous force per unit mass acting on fluid particle.
$\nu =$ fluid kinematic viscosity.

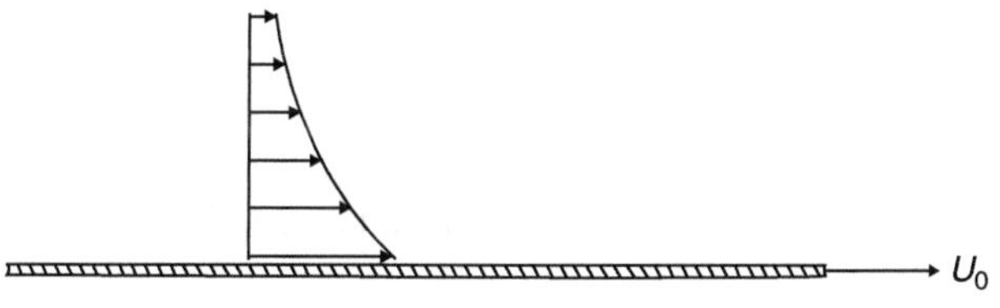

Fig. 4.4 Fluid velocities in vicinity of a plate suddenly set in motion

Note: Gravitational body forces are neglected and net pressure forces are zero.

Equation 4.32 is identical in the form of typical diffusion equation where temperature T is replaced by u while thermal diffusivity κ by ν. Refer to Sect. 4.2 of this chapter and Example 1 as well.

The initial and boundary conditions specified for Eq. 4.44 are

$$\lim_{t\to 0} u = 0 \text{ for all } \quad y > 0$$
$$u = U_0 \text{ for } \quad y = 0, \ t > 0$$
$$\lim_{y\to\infty} u = 0.$$

The term $\Theta(t)$ in Sect. 4.2 of this chapter is replaced by U_0 so that

$$u = U_0 F(\eta) \text{ where } \eta = \frac{y}{2\sqrt{\nu t}}.$$

The boundary and initial conditions on u are now applied to $F(\eta)$ and become

$$\lim_{\eta\to\infty} F(\eta) = 0$$
$$F(0) = 1 \text{ for } \quad t > 0.$$

Certain algebra manipulation and substitution of above equation will reduce the above PDE to ODE of the following form which is analogous to what we have in Eq. 4.3:

$$F'' + 2F' = 0. \tag{4.45}$$

The classical solution of Eq. 4.33 is given and readily found to be

$$F(\eta) = 1 = \frac{2}{\sqrt{\pi}} \int_0^\eta e^{-\eta^2} d\eta = \operatorname{erfc} \eta. \tag{4.46}$$

As before erfc η is the complimentary error function. Its value can be found from any mathematical table book. The solution for u is, therefore, given by

$$u = U_0 \operatorname{erfc} \eta. \tag{4.47}$$

Equation 4.35 is desired solution, but few points should be noted and they are [15] the following:

1. The method of solution required that Θ be strictly functioning of t. This means the problem might be chosen such that $\Theta(t) = c_2 t^n$ would not be a satisfactory solution. For example, we may consider the situation that the plate in the above example oscillated in a sinusoidal fashion. A natural choice for $\Theta(t)$ might then

be $\Theta(t) = U_0 \sin(mt)$. The method of solution as outlined would not be applicable for this choice of $\Theta(t)$.

2. The equation for $F(\eta)$, Eq. 4.33, is a second-order equation with two boundary conditions. Fortunately, the initial and boundary conditions $\lim_{t \to 0} u = 0$ and $\lim_{y \to \infty} u = 0$ combine into the single boundary condition on $F(\eta)$:

$$\lim_{\eta \to \infty} F(\eta) = 0.$$

The main advantage obtained here by transforming variables in the manner illustrated above is that a partial differential equation is reduced to an ordinary differential equation, and classical solution methods are employed in order to solve the transformed equation (Eq. 4.45).

It is good to know this technique is quite a powerful tool to solve nonlinear equations by reducing set of PDE to ODE where the elementary solution cannot be found as a closed form. However, points A and B above suggest certain important concerns:

- What general class of problems governed by the equation similar to diffusion equation can be solved in this manner?
- What boundary and initial conditions are required?
- Is it necessary that η and $\Theta(t)$ have the precise forms that were used in our example cases?
- What are other transformations, if any, which allows the PDE to be reduced in ODE form?

Further discussion can be found in Mattheij [14], Sachdev [3], Dresner [13], and Hansen [16].

Before we give the example provided by Hansen 16, we should point out the reason behind designating the solution obtained as a similarity solution in the following technique:

$$\frac{u}{U_0} = F(\eta).$$

With this assumption, thereby, the nondimensional velocity ratio u/U_0 is a function of the single variable η. A plot of η versus u/U_0 is shown in Fig. 4.5.

For any given time t_0, the parameter η can be replaced by a scalier multiple of y. Thus we see that if u/U_0 were plotted against y for various times, all velocity profiles would be "similar" in form. That is, the velocity profiles at various times differ only by coordinate scale changes, and this geometry property is a characteristic of similar solutions and is one reason for the name of similarity.

The physical problem under consideration is as follows: a semi-infinite flat plate is inserted in the steady uniform flow of an incompressible fluid with a small viscosity in such a way that it is aligned with the flow (see Fig. 4.4). The flow is assumed to be nonturbulent, and no variation in the flow or fluid properties is

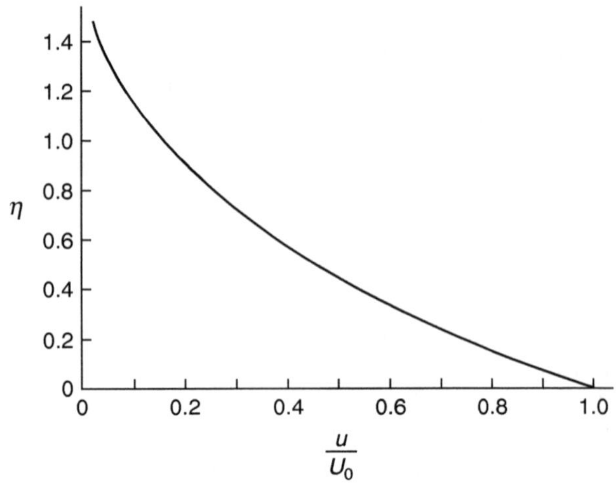

Fig. 4.5 A plot of η versus u/U_0

assumed to exist in a direction parallel to the plate leading edge; that is, two-dimensional flow prevails.

The effect of fluid viscosity is to cause the flow to adhere to the plate surface. It is also assumed that this effect is confined to rather thin layer of fluid near the plate surface and that outside of this layer the flow can be considered nonviscous. The fluid layer is called the boundary layer. Within the boundary layer, the fluid velocity varies from zero on the plate surface to the constant mainstream velocity at its outer edge.

Ludwig Prandtl formulated the mathematical theory of boundary-layer flow in 1904. Beginning with the momentum equations for a general viscous fluid flow, approximations were made which allow Prandtl to reduce the equations to the following single equation for the coordinate system and velocity components shown in Fig. 4.6:

$$u\frac{\partial u}{\partial x}+\upsilon\frac{\partial u}{\partial y}=\nu\frac{\partial^2 u}{\partial y^2}, \tag{4.48}$$

where u and υ are components of new coordinate system. See Fig. 4.7.

Note: In vicinity of the plate leading edge, the assumptions leading to Eq. 4.32 are no longer valid and the complete set of viscous flow equation may be applied.

The left side of Eq. 4.32 represents fluid acceleration, while the term on the right side represents force per unit mass arising fluid viscosity. A second equation can be obtained from the law of conservation of mass. The equation expressing this law for an incompressible flow is $\operatorname{div}\vec{V}=0$. Applied to the boundary-layer flow, this equation becomes

$$\frac{\partial u}{\partial x}+\frac{\partial \upsilon}{\partial y}=0. \tag{4.49}$$

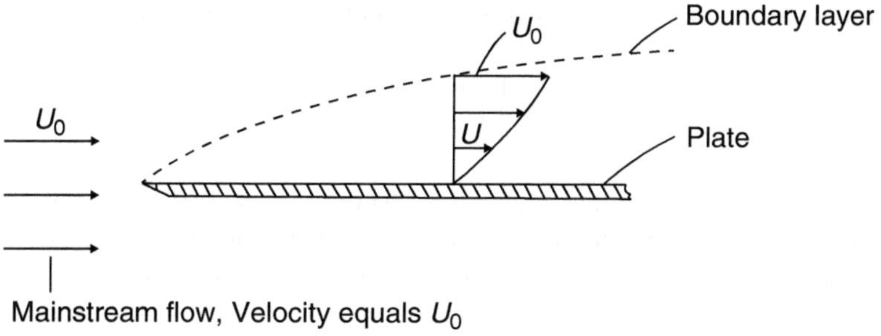

Fig. 4.6 Flat plate immersed in a uniform stream

Fig. 4.7 Coordinate system and velocity components for flow over a flat plate

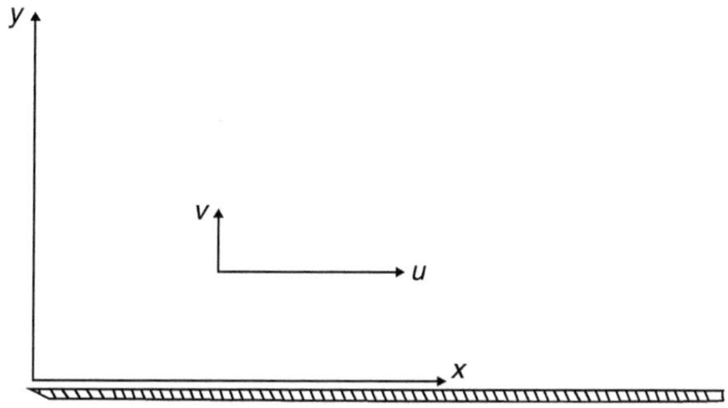

In dealing with two-dimensional, steady boundary-layer problems in general, the following boundary conditions are usually specified:

$$\begin{aligned} &u = 0 \quad v = 0 \text{ for} \quad y = 0 \\ &\lim_{x \to \infty} u = U_0(\text{Mainstream velocity}). \end{aligned} \tag{4.50}$$

In addition, a velocity profile $u = u(x_0, y)$ is required at some point $x = x_0$ [16]. Specially, we assume that the velocity component in the x-direction can be expressed as

$$\frac{u}{U_0} = \frac{dF}{d\eta} = F'(\eta) \tag{4.51}$$

where

$$\eta = y\sqrt{\frac{U_0}{\nu x}}. \tag{4.52}$$

Substitution of the expression for u into Eq. 4.49 and solving for υ give

$$\frac{\upsilon}{U_0} = \frac{1}{2}\sqrt{\frac{\nu}{U_0 x}}(\eta F' - F). \tag{4.53}$$

The Eq. 4.53 can be obtained under the assumption of $F(0) = 0$. Finally using Eq. 4.50 and substituting for expression u and υ result in the reduction of the equation to an ordinary differential equation in η, and we obtain 17:

$$\frac{FF''}{2} + F'' = 0. \tag{4.54}$$

The boundary condition associated with this equation will be given as three conditions that should be specified and can be derived from u and υ conditions presented in Eq. 4.50 which also must be satisfied by F. As $y = 0$ implies that $\eta = 0$ and as $\lim_{x\to\infty} \eta = \infty$, we can transform boundary conditions as follows:

$$\begin{aligned}
&u(x,0) = 0 \rightarrow F'(0) = 0\\
&u(x,0) = 0 \rightarrow F(0) = 0\\
&\lim_{x\to\infty} u = U_0 \rightarrow \lim_{F'\to\infty} F' = 1.
\end{aligned}$$

The solution of Eq. 4.38 can be obtained via standard numerical methods. The solution for $F'(\eta)$ is shown in Fig. 4.8.

It is important to recognize that in a general boundary-layer problem, the velocity profile $u(x_0, y)$ would be specified, but there is no way of taking such a requirement into account in solving the transformed Eq. 4.38. We thus encounter a possibly serious restriction in employing similarity transformation [15].

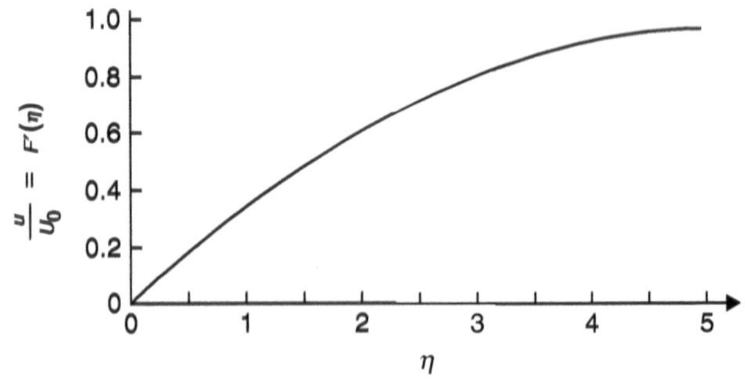

Fig. 4.8 u/U_0 versus η for flow over a flat plate

4.6 Solving First-Order Partial Differential Equations Using Similarity Method

First-order PDEs are usually classified as linear, quasilinear, or nonlinear. The first two types will be discussed in this section. Given a general form of first-order linear partial differential equation in two independent variables x and y as follows needs a solution for an application of interest that requires certain boundary and initial conditions be satisfied.

$$a(x,t)\frac{\partial u(x,y)}{\partial x}+b(x,t)\frac{\partial u(x,t)}{\partial y}=c(x,t)u+d(x,t), \tag{4.55}$$

where a, b, c, and d are functions of x and t only. A singular solution for given application for which the initial value problem fails to have a unique solution at some point on the solution. The set on which a solution is singular may be as small as a single point or as large as the full real line. This type of problem is known also as *singular Cauchy problem*. Solutions, which are singular in the sense that the initial value problem fails to have a unique solution, need not be singular functions. In some cases, the term singular solution is used to mean a solution at which there is a failure of uniqueness to the initial value problem at every point on the curve.

Singling out the variable t in most physical problem (i.e., diffusion equation), general form of Eq. 4.55 can be written in normal form of the following format:

$$\frac{\partial u}{\partial t}=F(x,t,u,u_x). \tag{4.56}$$

The general solution of first-order PDE involves an arbitrary function, but in solution of any application that involves this PDE, we are not interested in a general solution, but a solution that is subject to satisfying certain conditions such as imposed boundary and initial conditions applicable to the application and related PDE. These boundaries can be designated as BC and IC and more often are shown like this.

Few examples are shown here as part of demonstration of solution to a basic problem for a first-order PDEs which are holding a form of Eq. 4.56 as follows:

$$u_t-F(x,t,u,u_x)=0x\in R,\ t>0 \tag{4.57}$$

subject to *IC*

$$u(x,0)=u_0(x),\ x\in R, \tag{4.58}$$

where $u_0(x)$ is a given function in this case, and the interval of interest for x may be finite. This is a typical form of Cauchy problem, and it is a pure initial value problem and may be looked at as a signal or wave at $t=0$, so the initial signal or wave is a space distribution of u. The solution then may be obtained by plotting the graph of $u = u_0(x)$ in the xu-space. In this case, the PDE presented as Eq. 4.41 may be interpreted as the equation that describes the propagation of the wave as time increases. Sachdev [3] gives more details of discussion, and we use few of his and others' examples here for further clarifications. We have also discussed *perturbation theory* and *asymptotic methods* in Chap. 3 where demonstrated solutions of few examples were presented. As we said in that chapter, experience obtained during the analysis of various problems of applied mathematics, physics, and engineering with the use of asymptotic approaches indicates that perturbation techniques can be divided into two main groups, i.e., those devoted to the analysis of *regular problems* and *irregular problems* also known as *singular problem.* Now we present few examples here that are associated with some of the examples presented in Chap. 3. In case of regular perturbations, a solution $u_0(x)$ of the problem in space or interval of consideration can be found. Before we proceed with the examples, we will define the following:

Definition 1
Problem E_ε is understood as a regular perturbation [17]:

$$\sup_D \left|\left|u_\varepsilon(x) - u_0(x)\right|\right| \to 0 \text{ for } \quad \varepsilon \to 0. \tag{4.59}$$

Otherwise, problem E_ε is referred to as singular one.

In the formula above, $\sup_D f(x)$ denotes an upper branch of a function $f(x)$ in space D. In the case of regular perturbations, a solution $u_0(x)$ of the problem E_0 for small values of ε is closed to solution $u_\varepsilon(x)$ of the problem E_ε in whole space D. In the case of singular perturbations of $u_0(x)$ for small ε, $u_0(x)$ is not situated close to $u_\varepsilon(x)$, in some part of the space D this subspace is called *a* subspace of nonuniformity, illustrated in the following few examples:

Example 1 Find a solution to Cauchy problem of the following form [18]:

$$E_\varepsilon : \frac{du}{dx} = -u + \varepsilon x \quad x \in [0, 1] \quad u(0) = 1. \tag{4.60}$$

Solution The linear first-order differential Eq. 4.44 can be easily solved, and having satisfied the initial condition (IC) in it, one gets

$$u_\varepsilon(x) = (1 + \varepsilon)e^{-x} + \varepsilon(x - 1). \tag{4.61}$$

The associated problem is obtained from Eq. 4.60, taking E_0 and it reads

$$E_0 : \frac{du}{dx} = -u \quad x \in [0, 1] \quad u(0) = 1.$$

A solution to the above problem has the following form

$$u_0(x) = \mathrm{e}^{-x}. \tag{4.62}$$

Substituting Eqs. 4.61 and 4.62 into Eq. 4.59 of Definition 1 yields

$$\sup_{[0,1]} \left|\left|u_\varepsilon(x) - u_0(x)\right|\right| = c \max_{[0,1]} \left|c^{-x}\right|x - 1\left| > 0 \text{ for } \quad c > 0.$$

It means that according to Definition 1, the problem (Eq. 4.59) is a regular one, and the perturbation εx is a regular one as well.

Example 2 Find a solution to the following Cauchy problem [18]:

$$E_\varepsilon : \varepsilon \frac{du}{dx} = -u + x \quad x \in [0, 1] \quad U(0) = 1. \tag{4.63}$$

Solution The linear differential Eq. 4.46 is solved in a similar way to that of Eq. 4.43. Satisfying the initial condition Eq. 4.46 gives

$$u_\varepsilon(x) = (1 + \varepsilon)\mathrm{e}^{-x/\varepsilon} + x - \varepsilon \tag{4.64}$$

Note that the problem E_0 appeared in the following algebraic equation:

$$E_0 : 0 = -u + x \quad x \in [0, 1]. \tag{4.65}$$

Therefore, the initial condition can be omitted. Note also that

$$u_0(x) = x. \tag{4.66}$$

Substituting Eqs. 4.61 and 4.66 into Eq. 4.59 of Definition 1 for $\varepsilon \to 0$ yields

$$\sup_{[0,1]} \left|\left|u_\varepsilon(x) - u_0(x)\right|\right| = \max_{[0,1]} \left|(1 + \varepsilon)\mathrm{e}^{-x/\varepsilon} - \varepsilon\right| = 1.$$

According to Definition 1, problem Eq. 4.63 is singular, and, consequently, perturbation $\varepsilon \frac{du}{dx}$ is a singular one.

Drawings of solutions $u_0(x)$ versus $u_\varepsilon(x)$ for small $\varepsilon > 0$ are shown in Fig. 4.9.

Figure 4.9 clearly exhibits a characteristic feature of singular perturbations. It is visible that in the area $D(x) = \{0 \leq x \geq\}$, where problems E_ε and E_0 (above two examples and Eqs. 4.60 and 4.63) are defined, there is a subspace $[0, \delta), \delta = \delta(\varepsilon)$,

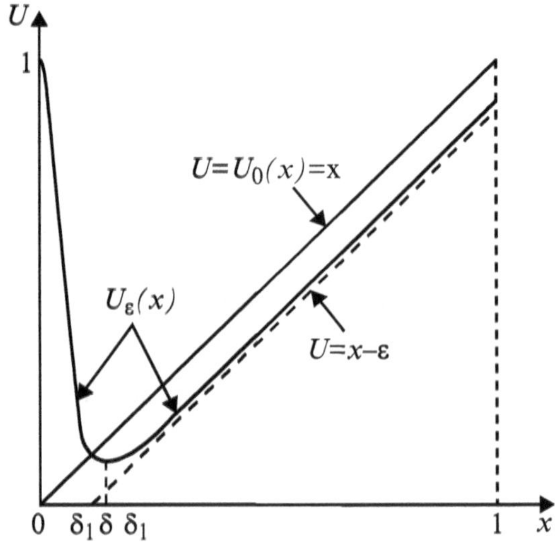

Fig. 4.9 Graphs of solution $u_0(x)$ versus $u_\varepsilon(x)$

where the solution $u_0(x)$ versus $u_\varepsilon(x)$, and a difference between them increases for $\varepsilon \to 0$. For further discussion and more examples, the reader should refer to Chap. 3 of the book by Awrejcewicz and Krysko [18].

Example 3 Find the solution of the following first-order partial differential equation [3]:

$$\frac{\partial u}{\partial t} + t^2 \frac{\partial u}{\partial x} = 0 \text{ for } \quad u(x,0) = f(x).$$

Solution It is clear that $\frac{du}{dt} = 0$ along the characteristic curves $\frac{dx}{dt} = t^2$. On integration we get $x = \frac{t^3}{3} + \xi$ so that

$$u = \text{constant on} x = \xi + \frac{t^3}{3}.$$

Therefore,

$$u(x,t) = u(\xi, 0) = f(\xi) = f\left(x - \frac{t^3}{3}\right).$$

The solution $u(x,t) = f\left(x - \frac{t^3}{3}\right)$ has a traveling wave from $f(\eta)$, $\eta = x - \frac{t^3}{3}$. The traveling wave moves with a nonconstant speed t^2 and a nonconstant acceleration $2t$.

The method of characteristics can also be applied to solve initial value problem (IVP) for a nonhomogeneous partial differential equation (PDE) of the form $u_t + c(x,t)u_x = f(x,t)$, $x \in R$, $t > 0$, and $u(x,0) = u_0(x)$.

Example 4 Find the solution of the following first-order partial differential equation [3]:

$$\frac{\partial u}{\partial t} + c\frac{\partial u}{\partial x} = \mathrm{e}^{-3x} \text{ for } u(x,0) = f(x).$$

Solution We note that

$$\frac{du}{dt} = \mathrm{e}^{-3x} \text{ along } \frac{dx}{dt} = c.$$

This pair of ODEs can be solved subject to the IC $x = \xi$, $u = f(\xi)$ at $t = 0$. We get

$$x = ct + \xi.$$

On integration we have

$$u(x,t) = \frac{\mathrm{e}^{-3ct}}{-3c}\mathrm{e}^{-3\xi} + g(\xi),$$

where g is the function of integration. Applying the IC we get

$$g(\xi) = \frac{\mathrm{e}^{-3\xi}}{3c} + f(\xi).$$

Thus,

$$\begin{aligned} u(x,t) &= \frac{\mathrm{e}^{-3\xi}}{3c}\left(1 - \mathrm{e}^{-3ct}\right) + f(\xi) \\ &= \frac{\mathrm{e}^{-3(x-ct)}}{3c}\left(1 - \mathrm{e}^{-3ct}\right) + f(x - ct). \end{aligned}$$

The solution here is of the similarity form $u(x,t) = \alpha(x,t) + \beta(\eta)$, where $\eta = x - ct$ is the similarity variable, a linear combination of the independent variable x and t.

Example 5 Find the solution of the following first-order partial differential equation [3]:

$$\frac{\partial u}{\partial t} + x\frac{\partial u}{\partial x} = t \text{ for } \quad u(x,0) = f(x).$$

Solution Here

$$\frac{du}{dt} = t \text{ along} \frac{dx}{dt} = x,$$

which on integration yields

$$x = \xi e^t$$

and

$$u(x,t) = \frac{t^2}{2} + g(\xi).$$

At $t = 0$, $x = \xi$, and $u = f(\xi)$; therefore, $g(\xi) = f(\xi)$. Thus

$$u = \frac{t^2}{2} + f(\xi) = \frac{t^2}{2} + f(xe^{-t}).$$

The solution here has the similarity form

$$u = \alpha(x,t) + \beta(\eta),$$

where $\eta = xe^{-t}$ is the similarity variable.

Example 6 Find the solution of the following first-order partial differential equation [3]:

$$x\frac{\partial u}{\partial x} + (x^2 + y)\frac{\partial u}{\partial y} + \left(\frac{y}{x} - x\right)u = 1.$$

Solution The characteristics are given by

$$\frac{dx}{dt} = x\frac{dy}{dt} = x^2 + t\frac{du}{dt} + \left(\frac{y}{x} - x\right)u = 1,$$

the first two of which give the locus in the (x, y) plane, the so-called traces

$$\frac{dy}{dx} = x + \frac{y}{x},$$

which on integration become

$$\frac{y}{x} - x = \text{constant}.$$

It is often easier to find the general solution of the PDE by introducing the variable describing the trace curves as a new independent variable: $\phi = \frac{y}{x} - x$. The given PDE then becomes

$$x\left(\frac{\partial u}{\partial x}\right)_{\phi} + \phi u = 1,$$

which on integration with respect to ϕ gives

$$u = \phi^{-1} + x^{-\phi} f(\phi),$$

where f is an arbitrary function of ϕ.

4.6.1 *Solving Quasilinear Partial Differential Equations of First-Order Using Similarity*

Before we concentrate and restrict our attention to first-order quasilinear partial differential equations, we need to introduce certain definitions that were given as a series of lectures as class note by Professor H. M. Atassi of University of Notre Dame, Department of Aerospace and Mechanical Engineering on the subject of *Quasilinear Partial Differential Equations* as follows:

- **Definition 1:** An equation containing partial derivatives of the unknown function u is said to be an n-th order equation if it contains at least one n-th order derivative, but contains no derivative of order higher than n.
- **Definition 2:** A partial differential equation is said to be linear if it is linear with respect to the unknown function and its derivatives appear in it.
- **Definition 3:** A partial differential equation is said to be quasilinear if it is linear with respect to all the highest-order derivatives of the unknown function.

Example 1 The equation

$$\frac{\partial^2 u}{\partial x^2} + a(x, y)\frac{\partial^2 u}{\partial y^2} - 2u = 0$$

is a second-order linear partial differential equation. However, the following equation

$$\frac{\partial u}{\partial x}\frac{\partial^2 u}{\partial x^2}+\frac{\partial u}{\partial y}\frac{\partial^2 u}{\partial y^2}+u^2=0$$

is a second-order quasilinear partial differential equation. Finally, the equation

$$\left(\frac{\partial u}{\partial x}\right)^2+\left(\frac{\partial u}{\partial y}\right)^2-u=0$$

is a first-order partial differential equation which is neither linear nor quasilinear.

- A solution of a partial differential equation is any function that, when substituted for the unknown function in the equation, reduces the equation to an identity in the unknown variables.

Example 2 Let us consider the one-dimensional wave equation

$$\left(\frac{\partial^2}{\partial t^2}-c^2\frac{\partial^2}{\partial x^2}\right)u=0.$$

It is well known, and will be shown later, that the general solution of this equation can be cast as

$$u=f(x-ct)+g(x+ct),$$

where f and g are arbitrary twice differentiable functions of the single variable $\xi = x-ct$ and $\eta = x+ct$, respectively. It is easy to see, using the chain rule, that

$$\frac{\partial^2 u}{\partial t^2}=c^2\left(\frac{d^2 f}{dx^2}+\frac{d^2 g}{dy^2}\right)$$
$$\frac{\partial^2 u}{\partial x^2}=\frac{d^2 f}{dx^2}+\frac{d^2 g}{dy^2}.$$

Substitution into the wave equation leads to an identity.

- **Definition 5:** Let $\phi(\vec{x})$ be a function of the vector $\vec{x}=(x_1,x_2,\cdots,x_n)$ having first-order derivatives. Then the vector in R^n

$$(D_1,D_2,\cdots,D_n)\phi,$$

where $D_i=\frac{\partial}{\partial x_i}$ is called the gradient of ϕ and usually denoted as "grad ϕ" or "$\nabla\phi$."

- **Definition 6:** Let $\vec{v}$ be a unit vector in R^n and v measures the distance on $\vec{v}$. Then the limit

$$\frac{d\phi(\vec{x})}{dv} = \lim_{\Delta t \to 0} \frac{\phi(\vec{x} + \vec{v}\Delta t) - \phi(\vec{x})}{\Delta t}.$$

If it exists, it is called the derivative of ϕ in the $\vec{v}$ direction. It is easy to show that

$$\frac{d\phi(\vec{x})}{dv} = \nabla\phi(\vec{x}) \cdot \vec{v} = (v_1 D_1, v_2 D_2, \cdots, v_n D_n)\phi(\vec{x}),$$

where $\vec{v} = (v_1, v_2, \cdots, v_n)$.

Example 3

1. In R^2, $\vec{x} = (x, y)$. Then

$$\text{grad}\phi(x,y) = (D_x, D_y)\phi = \left(\frac{\partial \phi}{\partial x}, \frac{\partial \phi}{\partial y}\right).$$

2. Let $\phi(x, y, z)$ be a function having continuous first-order derivative. The equation

$$\phi(x,y,z) = c,$$

where $c \in R$, represents a surface S. As we move along a path $\gamma(s)$ on the surface S.

$$\frac{d\phi}{ds} = \frac{\partial \phi}{\partial x}\frac{dx}{ds} + \frac{\partial \phi}{\partial y}\frac{dy}{ds} + \frac{\partial \phi}{\partial z}\frac{dz}{ds} = 0$$

$$\frac{d\phi}{ds} = \text{grad}\phi \cdot \left(\frac{dx}{ds}, \frac{dy}{ds}, \frac{dz}{ds}\right) = 0.$$

For an infinitesimal change in ds, the vector $(dx/ds, dy/ds, dz/ds)$ is in the plane tangent to the surface, at the point $M(x, y, z)$. Consequently, gradϕ is orthogonal to the surface $\phi = c$.

Now that we got all above definitions out of our way, we now restrict our exposition to first-order quasilinear partial differential equations (FOQPDE) with two variables, since this case affords a real geometric interpretation. However, the treatment can be extended without difficulty to higher-order spaces. The general form of FOQPDE with two independent variables is

$$a(x,y,y)\frac{\partial u}{\partial x} + b(x,y,u)\frac{\partial u}{\partial y} = c(x,y,u), \tag{4.67}$$

where a, b, and c are continuous functions with respect to the three variables x, y, u. Let $u = u(x, y)$ be a solution to Eq. 4.67. If we identify u with the third coordinate z in R^3, then $u = u(x, y)$ represents a surface S. The direction of the normal to S is the vector $\left(u_x, u_y, -1\right)$, where u_x, u_y are shorthand notations of $\frac{\partial u}{\partial y}$, respectively. On the other hand, Eq. 4.67 can be written as the inner product

$$\left(u_x, u_y, -1\right) \cdot (a, b, c) = 0. \tag{4.68}$$

Thus, (a, b, c) is perpendicular to the normal to S and, consequently, must lie in the plane tangent to S. Let us consider a path $\gamma(s)$ on S. The rate of variation of u as we move along γ is

$$\frac{du}{ds} = \frac{\partial u}{\partial x}\frac{dx}{ds} + \frac{\partial u}{\partial y}\frac{dy}{ds}. \tag{4.69}$$

The vector $(dx/ds, dy/ds, du/ds)$ is naturally the tangent to the curve $\gamma(s)$. Equation 4.69 can be rewritten as the inner product

$$\left(u_x, u_y, -1\right) \cdot (dx/ds, dy/ds, du/ds) = 0. \tag{4.70}$$

Comparing Eqs. 4.68 and 4.70, we see that there is a particular family of curves on the surface S, defined by

$$\frac{dx/ds}{a} = \frac{dy/ds}{b} = \frac{du/ds}{c}. \tag{4.71}$$

These curves are called characteristics and will be denoted by $C(s)$ or simply C. We note that there are actually only two independent equations in the system (Eq. 4.55); therefore, its solutions comprise in all a two-parameter family of curves in space.

Theorem 1 *Any one-parameter subset of the characteristics generates a solution of the first-order quasilinear partial differential Eq.* 4.51.

Proof Let $u = u(x, y)$ be the surface generated by a one-parameter family $C(s)$ of the characteristics whose differential equations are 4.54. By taking the rate of variation of u along a characteristic curve $C(s)$, we get

$$\frac{du}{ds} = u_x\frac{dx}{ds} + u_y\frac{dy}{ds}$$

or

$$\left(u_x, u_y, -1\right) \cdot (dx/ds, dy/ds, du/ds) = 0. \tag{4.72}$$

However, Eq. 4.71 state that the vectors $(dx/ds, dy/ds, du/ds)$ and (a, b, c) are collinear. Therefore,

$$\left(u_x, u_y, -1\right) \cdot (a, b, c) = 0$$

or

$$au_x + bu_y = c.$$

In addition, we have proven the case and done with it.

Corollary 1 *The general solution to Eq.* 4.71 *is defined by a single relation between two arbitrary constants occurring in the general solution of the system of ordinary differential equations:*

$$\frac{(dx/ds)}{a} = \frac{(dy/ds)}{b} = \frac{(du/ds)}{c},$$

in other words, by any arbitrary function of one independent variable.

Example 4 Consider the first-order partial differential equation:

$$\frac{\partial u}{\partial t} + u\frac{\partial u}{\partial x} = 0.$$

The characteristic is defined by

$$\frac{(dt/ds)}{1} = \frac{(dx/ds)}{u} = \frac{(du/ds)}{0}.$$

The last equation gives immediately $u = k_1$. Since u is constant along a given characteristic, then the first equation can be integrated immediately:

$$x - k_1 t = k_2.$$

The general solution is then

$$k_1 = f(k_2),$$

where f is an arbitrary function of the single variable k_2. Substituting k_1 and k_2 by their expressions, we finally have

$$u = f(x - ut).$$

Example 5 Consider the equation

$$3\frac{\partial u}{\partial x} = 7\frac{\partial u}{\partial y} = 0.$$

The characteristics are solution of the system

$$\frac{(dx/ds)}{3} = \frac{(dy/ds)}{-7} = \frac{(du/ds)}{0}.$$

By integration, we get

$$u = k_1$$
$$3y + 7x = k_2.$$

The characteristics are straight lines intersection of the two families of planes defined by these equations. Any arbitrary relation between k_1 and k_2 is a solution. The general solution is then

$$u = f(3y + 7x),$$

where f is an arbitrary function of the one independent variable $3y + 7x$.

Example 6. Consider the equation

$$y\frac{\partial u}{\partial x} - x\frac{\partial u}{\partial y} = 0.$$

The characteristics are solution of the system

$$\frac{(dx/ds)}{y} = \frac{(dy/ds)}{-x} = \frac{(du/ds)}{0}.$$

By integration, we get

$$u = k_1$$
$$x^2 + y^2 = k_2.$$

The characteristics are circles located in the plane $u = k_1$. The general solution is

$$u = f(x^2 + y^2),$$

where f is an arbitrary function of the independent variable $x^2 + y^2$. Geometrically, the general solution is any surface of revolution around the u-axis.

Remark 1: The previous examples illustrate the application of the theory of characteristics to find the general solution to a first-order quasilinear partial differential equation. Given the simple forms of these examples, the student could become suspicious that in the general case, it will not be possible to get a closed analytical form of the solution. A careful examination of system (Eq. 4.55) success to convince you that, in the general case, i.e., when a, b, and c are arbitrary functions of x, y, and u, the system of ordinary differential equations (Eq. 4.55) cannot be integrated analytically. Nevertheless, for a given initial or boundary value problem, system (Eq. 4.55) provides a numerical solution.

Remark 2: When Eq. 4.67 is a linear homogeneous partial differential equation,

$$a(x,y)u_x + b(x,y)u_y = c(x,y)u. \tag{4.73}$$

The solutions form a vector space which we call the solution space. This solution space is naturally the null space of the linear operator:

$$a(x,y)D_x + b(x,y)D_y - c(x,y). \tag{4.74}$$

It should be pointed out that since the general solution to Eq. 4.73 is any arbitrary function of one independent variable, there is a nondependent numerable infinity of independent solutions to Eq. 4.73. Therefore, the dimension of null space of the operator (Eq. 4.74) is infinity. This result appears to be in sharp contrast with what we know about the first-order ordinary linear differential operators where the null space is one-dimensional. One may also add that this augurs the difficulties we shall encounter in the study of partial differential operators.

4.6.2 The Boundary Value Problem for a First-Order Partial Differential Equation

The theory of characteristics enables us to define the solution to FOPDE (Eq. 4.67) as surfaces generated by the characteristic curves defined by the ordinary differential equations (Eq. 4.71). However, a physical problem is not uniquely specified if we simply give the differential equation, which the solution must satisfy, for, as we have seen, there are an infinite number of solutions of every equation. In order to make the problem a definite one, with a unique answer, we must pick out of the mass of possible solutions, the one that has certain definite properties along definite boundary surfaces. These properties represent the boundary conditions, which the solution must satisfy. The first fact, which we must notice, is that we cannot try to make the solutions of a given equation satisfy any sort of boundary conditions, for there is a definite set of boundary conditions, which will give nonunique or impossible answers. The study of the proper boundary conditions to be specified on definite boundary curves or surfaces is often termed the Cauchy problem, in honor of the French mathematician who laid the foundations to our present knowledge in this area.

4.6.3 Statement of the Cauchy Problem for First-Order Partial Differential Equation

Let $\gamma(t)$ be a curve in region R of the x, y plane, where the value of the function $u(x, y)$ which satisfies Eq. 4.67 is specified as $u(t)$. What are the conditions to be satisfied by $\gamma(t)$ and $u(t)$, in order that the boundary value problem so defined has a unique solution?

Consider an initial curve $\gamma(t)$ in a region R

$$\begin{aligned} \xi &= \xi(t) \\ \eta &= \eta(t). \end{aligned} \tag{4.75}$$

Since $u(x, y)$ is specified on $\gamma(t)$ as

$$u(x, y) = u(t), \tag{4.76}$$

then the three functions are

$$\xi = \xi(t) \;\; \eta = \eta(t) \;\; u = u(t). \tag{4.77}$$

Define a curve Γ in space whose projection over the $x - y$ plane is γ. By a proper choice of the parameters s, the equations of the characteristics are

$$\frac{dx}{ds} = a\frac{dy}{ds} = b\frac{du}{ds} = c. \tag{4.78}$$

The characteristic curve passing by a point $M(t) \in \Gamma$ is then defined by the following equations:

$$x = x(s, t) \; y = y(s, t) \; u = u(s, t). \tag{4.79}$$

Equation 4.79 represents the surface S in a parametric form. When t is kept constant, we move along a characteristic curve $C(s)$. For a given value of s, for example, s_0, we move on the surface S along the curve Γ (Fig. 4.10).

Let us now assume that the initial curve Γ was a characteristic curve C. It is then obvious that Eq. 4.75 satisfies system (Eq. 4.78), and there will be no surface solution generated this way. We conclude that in order to generate a surface solution, Γ should not be a characteristic curve. Mathematically, this means that

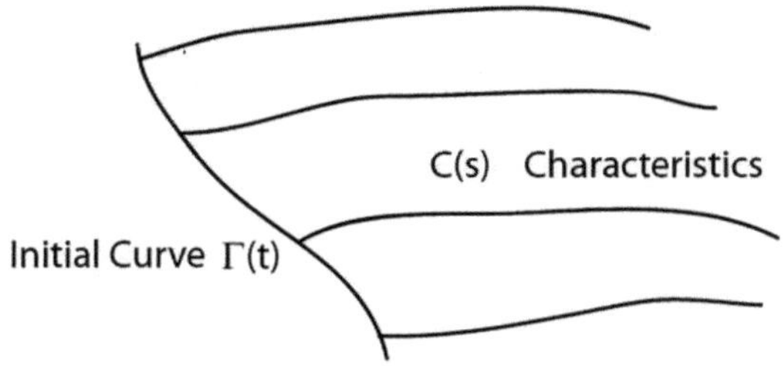

Fig. 4.10 Solution of the boundary value problem of a first-order partial differential equation by a one-parameter family of characteristics

the two vectors $(d\xi/dt, d\eta/dt, du/dt)$ and (a, b, c) must be linearly independent. A sufficient condition to insure this linear independence is that the determinant

$$J = \begin{vmatrix} \frac{d\xi}{dt} & \frac{d\eta}{dt} \\ a & b \end{vmatrix}. \tag{4.80}$$

Never vanishes on Γ.

An alternative form to the condition (Eq. 4.80) can be obtained from Eq. 4.79. The vectors tangent to C and Γ are, respectively, (x_s, y_s, u_s) and (x_t, y_t, u_t). A sufficient condition for their independence is that the Jacobian

$$J = \begin{vmatrix} x_s & y_s \\ x_t & y_t \end{vmatrix}.$$

Never vanishes on Γ. We are then led to the following theorem:

Theorem 2 *The solution $u(x, y)$ may be freely specified along a curve γ in R, and the resulting specification determines a unique solution of (Eq.* 4.51*) if and only if γ intersects the projection on the $x - y$ plane of each characteristic curve exactly once.*

Example 7 Find the general solution of the following partial differential equation:

$$(t + u)\frac{\partial u(x, t)}{\partial x} + t\frac{\partial u(x, t)}{\partial t} = x - t.$$

Also find the integral surface containing the curve $t = 1$, $-\infty < x < +\infty$.

Solution The characteristic of the given PDE are identified by

$$\frac{dx}{t + u} = \frac{dt}{t} = \frac{du}{x - t}.$$

It is easy to see that

$$\frac{d(x+u)}{x+u} = \frac{dt}{t}.$$

On integration we have

$$\frac{x+u}{t} = c_1,$$

where c_1 is a constant. Again

$$\frac{d(x-t)}{u} = \frac{du}{x-t}$$

implying

$$(x-t)^2 - u^2 = c_2,$$

where c_2 is another constant.

The general solution, therefore, is

$$(x-t)^2 - u^2 = f\left(\frac{x+u}{t}\right).$$

If the integral surface contains the given curve $t = 1$, $u = 1 + x$, we have

$$(x-t)^2 - (1+x)^2 = f(1+2x)$$

or

$$f(1+2x) = -4x$$

implying that

$$f(z) = -2(z-1)$$

and also

$$f\left(\frac{x+u}{t}\right) = -2\left(\frac{x+u}{t} - 1\right).$$

The solution therefore is

$$(x-t)^2 - u^2 = -\frac{2}{t}(x+u-t).$$

Solving for u, we have

$$u = \frac{1}{t} \pm \left(x - t + \frac{t}{t}\right).$$

The condition $u = 1 + x$ when $t = 1$ is satisfied only if we take the positive sign. Thus, the solution of the initial value problem (IVP) is

$$u = \frac{2}{t} + x - t.$$

Clearly, the solution is defined only for $t > 0$.

While the general solution is quite implicit, the solution of IVP has the form $u = f(t) + g(\eta)$, $\eta = x - t$, and may be found by similarity methods.

Example 8 Find the general solution of the following partial differential equation:

$$(t^2 - u^2)\frac{\partial u(x,t)}{\partial x} - xt\frac{\partial (x,t)}{\partial t} = xu.$$

Also find the integral surface containing the curve $x = t = u$, $x > 0$.

Solution The characteristics of the given partial differential equation (PDE) are

$$\frac{dx}{t^2 - u^2} = \frac{dt}{-xt} = \frac{du}{xu}.$$

A first integral obtained from the second pair is $\phi(x,t,u) \equiv ut = c_1$. Each of the above ratio is equal to

$$\frac{xdx + tdt + udu}{x(t^2 - u^2) + t(-xt) + u(xu)} = \frac{xdx + tdt + udu}{0}.$$

Therefore, a second integral is $\psi(x,t,u) \equiv x^2 + t^2 + u^2 = c_2$ to say. The general solution, therefore, is $\phi = f(\psi)$, that is

$$ut = f\left(x^2 + t^2 + u^2\right).$$

Applying the initial condition $x = t = u$, we get

$$x^2 = f\left(3x^2\right)$$

giving

$$f(z) = \frac{z}{3}.$$

Therefore, we get the special solution satisfying the initial condition (IC) as

$$ut = \frac{x^2 + t^2 + u^2}{3}.$$

Solving the quadratic in u, we find that

$$u = \frac{3t - (5t^2 - 4x^2)^{1/2}}{2},$$

the root with the negative sign satisfying the given conditions. Here, again, the general solution is rather implicit. The special solution satisfying given IC may be obtained by the similarity approach.

4.7 Exact Similarity Solutions on Nonlinear Partial Differential Equations

Nonlinear partial differential equations (PDEs) generally do not present exact linearization, and then one should deal with them directly either using similarity method to find solutions or their more generalized forms. Exact similarity solutions from early time were intuitively soughed in the form $\vec{u} = t^m \vec{f}(xt^n)$ when the PDEs involved two independent variables such as m and n that were found either by dimensional analysis argument as Pi theorem or direct substitution so that PDEs reduced to ODEs (see Sect. 4.3). This class was fully identified by the use of Lie group methods as described in this chapter along with associated examples or intuitively by the direct similarity approach using infinitesimal transformation to identify the similarity form of the solution [3]. The Lie group method of infinitesimal transformations is the classical method of finding symmetry reductions of PDEs.

An alternative way based on group techniques that could be used to obtain the same exact solutions of the PDEs solved in this chapter is discussed. These techniques lead to solutions in special forms and are obtained by exploiting the symmetries of the original equation. Symmetry techniques provide a method for getting exact solutions of a certain PDE in terms of solutions of lower dimensional equations. An advantage of these techniques is that they are applicable to all PDEs, irrespective whether the equations are integrable.

To apply this method to a second-order equation of the form [19]

$$G(x, t, u, u_x, u_t, u_{xx}) = 0, \tag{4.81}$$

where x and t are the independent variables, one considers the one-parameter (ε) Lie group of infinitesimal transformations in (x, t, u), given by

$$\begin{aligned}\bar{x} &= x + \varepsilon\xi(x,t,u) + O(\varepsilon^2)\\ \bar{t} &= t + \varepsilon\tau(x,t,u) + O(\varepsilon^2)\\ \bar{u} &= u + \varepsilon\eta(x,t,u) + O(\varepsilon^2).\end{aligned} \tag{4.82}$$

The requirement that Eq. 4.81 is invariant under this transformation yields an overdetermined, linear system of equations for the infinitesimals $\xi(x,t,u)$, $\tau(x,t,u)$, and $\eta(x,t,u)$. The corresponding Lie algebra is realized by vector fields of the form

$$X = \tau(x,t,u)\frac{\partial}{\partial t} + \xi(x,t,u)\frac{\partial}{\partial x} + \eta(x,t,u)\frac{\partial}{\partial u}. \tag{4.83}$$

Similarity reductions are then obtained by solving the characteristic equations

$$\frac{dx}{\xi(x,t,u)} = \frac{dt}{\tau(x,t,u)} = \frac{du}{\eta(x,t,u)} \tag{4.84}$$

or, equivalently, the invariant surface condition

$$\xi(x,t,u)u_x + \tau(x,t,u)u_t + \eta(x,t,u) = 0. \tag{4.85}$$

Solving the system of the overdetermined equations for ξ, τ, and η involves large amounts of algebra and calculus and requires the use of symbolic manipulation programs. Many programs have been developed to facilitate these calculations. There is good discussion summary of classical and nonclassical methods to deal with direct method of exact solution of PDEs given by Nuseir [19].

The power of computer these days provides programs that are available to lead directly to the form, thus mitigating the considerable effort involved in the procedure. Numerical approach in the case of direct similarity and finding required solutions no longer requires knowledge of the group invariance property of partial differential equations (PDEs). Clarkson and Kruskal [20], Nuseir [19], and VU et al. offer some of these types of programming and solutions [21].

Systems of differential equations occur often in many theoretical and applied areas. In many cases, exact solutions are required as numerical methods are not appropriate or applicable. Indeed, exact solutions of systems of partial differential equations arising in fluid dynamics, continuum mechanics, and general relativity are of considerable value for the light they shed into extreme cases, which are not susceptible to numerical treatments. One important source of exact solutions to differential equations is the application of the group theoretic method of Lie. Such solutions found by Lie's method are called invariant solutions. Essential to this approach is the need to solve overdetermined systems of "determining equations," which consist of coupled, linear, homogeneous, partial differential equations. Typically, such systems vary between ten to several hundred equations. Clearly in the case of sets of equations consisting of about 100 equations or more, the prospect of finding solutions to such systems with just pencil and paper would certainly be quite challenging.

Fakhar [22] has investigated and published his paper on exact solutions for an unsteady flow of an incompressible fluid of third-grade boundary by an infinite porous plate and has obtained velocity component. He has performed Lie symmetry analysis to obtain the solution and symmetries of translational type. The mechanics of nonlinear fluids present a special challenge to engineers, physicists, and mathematicians since the nonlinearity can manifest itself in a variety of ways [22]. One of the simplest ways in which the viscoelastic fluids have been classified by Rivlin and Ericksen [23] and Truesdell and Noll [24]. They present constitutive relations for the stress tensor as a function of the symmetric part of the velocity gradient and its higher (total) derivatives.

The linear PDEs with variable coefficients are not much easier than the nonlinear ones as far as their explicit solutions are concerned; the only major advantage is they can use the principle of linear superposition. Sachdev discusses about other complicating factor; even when we linearize a nonlinear PDE, the corresponding initial/boundary conditions generally transform in such a cumbersome way that the solution of the exactly linearized problem is rendered difficult. For example, problems of one-dimensional, time-dependent where motion of an ideal compressible isentropic gas in the hodograph plane is under study or even two-dimensional, isentropic steady-flow equations where water waves up a uniformly sloping beach is dealt with are good examples of this scenario. Sachdev [3] in Chap. 5 of his book in Sects. 5.5–5.7 provides good examples of the above cases.

Note if the linear PDEs have constant coefficients, and the relevant ICs/BCs are not too complicated, the integral transform techniques (i.e., Laplace, Fourier, etc.) usually suffice to solve them. This is a consequence of the principle of linear superposition. However, if the linear PDEs have variable coefficients, integral transform techniques can still be employed, but finding the inverse transform is not always feasible and remains to be a challenge of advanced mathematics. One may still use asymptotic method to solicit some solution to look at behavior of the dependent variable for a particular range of one of the independent variables. See next section for brief approach of this method, and further knowledge can be found in different literatures and related books about this topic and method.

4.8 Asymptotic Solutions by Balancing Arguments

For nonlinear partial differential or ordinary equation problems where obtaining exact solution is not achievable, one may look into asymptotic solution for large time or distance considering the given BCs and ICs. For this purpose a perturbation theory can help to sift from the full equation those terms which balance in terms of power of the limiting variables, the remaining terms being small in comparison of the limit, and the effect of neglected terms may be incorporated in the next order. Sachdev [3] is considering and investigating behavior of the following equation:

$$y^2 y'' = -\frac{1}{3} \tag{4.86}$$

as the independent variable $x \to \infty$. One possible solution is a quadratic form for y as a function of x and a small correction term, which may include the effect of the term $-\frac{1}{3}$ on the right-hand side of Eq. 4.86 as follows:

$$y(x) \sim ax^2 + bx + c + \varepsilon(x). \tag{4.87}$$

Substituting Eq. 4.87 into Eq. 4.70 and assuming that $\xi(x)$ is small as $x \to \infty$, we have

$$a^2 x^4 \varepsilon''' \sim -\frac{1}{3} \text{ as } x \to \infty. \tag{4.88}$$

Integrating Eq. 4.88 yields

$$\varepsilon \sim \frac{1}{18a^2 x} \quad x \to \infty, \tag{4.89}$$

which easily can be seen as $x \to \infty$. With the correction term (Eq. 4.89) put in (Eq. 4.87), one may justify a general solution and assume

$$y(x) \sim ax^2 + bx + c + \frac{d}{x} + \frac{e}{x^2} + \frac{f}{x^3} + \cdots \tag{4.90}$$

Substituting Eq. 4.90 into Eq. 4.86 and comparing like powers of x on both sides, it is found that

$$d = \frac{1}{18a^2} \quad e = \frac{b}{36a^3} \quad f = \frac{3b^2 - 2ac}{180a^4} \quad \text{etc.} \tag{4.91}$$

in addition, therefore, the solution takes the form of

$$y(x) \sim ax^2 + bx + c + \frac{1}{18a^2} - \frac{b}{36a^3} + \frac{3b^2 - 2ac}{180a^4} + \cdots \text{ as } x \to \infty. \tag{4.92}$$

The behavior of the series in Eq. 4.92 for large x maybe also investigated analytically or numerically as well. This series is clearly singular for $a = 0$ (b and c are other arbitrary constants), suggesting that there is another asymptotic behavior which covers that this case exists [3]. If we attempt

$$y(x) \sim Ax^{\alpha}, x \to \infty \tag{4.93}$$

for Eq. 4.86, we find that

$$A^3\alpha(\alpha-1)(\alpha-2)x^{3\alpha-3} \sim -\frac{1}{3}\ x \to \infty \tag{4.94}$$

giving $\alpha = 1$, which, however, leads to a contradiction. In such circumstances it is usual to attempt

$$y(x) \sim Ax(\ln x)^{\alpha}, \tag{4.95}$$

which may correct the choice Ax. Putting Eq. 4.95 into Eq. 4.86, we find that

$$A^2x^2(\ln x)^{2\alpha}\left[-A\alpha\frac{(\ln x)^{\alpha-1}}{x^2} + \frac{A\alpha(\alpha-1)(\alpha-2)(\ln x)^{\alpha-3}}{x^2}\right] \sim -\frac{1}{3}. \tag{4.96}$$

Neglecting $(\ln x)^{\alpha-3}$ in comparison with $(\ln x)^{\alpha-1}$ as $x \to \infty$, we find that $-A^3\alpha(\ln x)^{3\alpha-1} \sim -\frac{1}{3}$ as $x \to \infty$. We thus have the choice $\alpha = \frac{1}{3}$ and $A = 1$. To lowest order, we have

$$y(x) \sim x(\ln x)^{1/3}, x \to \infty. \tag{4.97}$$

The next step is to attempt to improve upon Eq. 4.97. One "plausible" choice is to write a descending power series in $\ln x$:

$$y(x) \sim x(\ln x)^{1/3}\left[1 + \frac{A}{(\ln x)} + \frac{B}{(\ln x)^2} + \frac{C}{(\ln x)^3} + \cdots\right]. \tag{4.98}$$

If we substitute Eq. 4.98 into Eq. 4.86 and equate various power of $(\ln x)$ on both sides, it turns out that A is arbitrary and $B = -\frac{10}{27} - A^2$, $C = \frac{50}{27}, A + \frac{5}{3}A^3$, etc. Thus, Eq. 4.98 becomes

$$y(x) \sim x(\ln x)^{1/3}\left[1 + \frac{A}{\ln x} - \frac{A^2 + \frac{10}{27}}{(\ln x)^2} + \frac{\frac{5}{3}A^3 + \frac{50}{27}A}{(\ln x)^3} + \cdots\right], x \to \infty. \tag{4.99}$$

Actually, Eq. 4.86 can be solved in a closed form following a sequence of perfectly logical transformations [3], but the final solution is so implicit that it has little practical use. Sachdev [3] argues that on the other hand, the behaviors of Eqs. 4.86 and 4.99 for $x \to \infty$ when combined with appropriate numerical solution can provide useful information about the structure of the solution for large x. See also Bender and Orszag [25].

Note

that any form of Eq. 4.93 is very similar approach to analyze the energy in a high intense implosion, which is typical self-similar approach of second kind (see Chap. 1, Sects. 1.10 and 1.16).

Also per Barenblatt [26] definition here, "The term "scaling" denotes a seemingly very simple thing: a power-law relationship between certain variables y and x of the form

$$y = Ax^{\alpha},$$

where A, α are constant. Such relations often appear in mathematical modeling of various phenomena, not only in physics, but also in biology, economics, and engineering. Scaling laws are not merely some special simple cases of more general relations. They never appear by accident. Scaling laws always reveal a very important property of the phenomenon under consideration its self-similarity. Self-similar means reproducing itself on different time and space scales."

More interesting, work on asymptotic of nonlinear ODE by balancing argument is argued and was presented by Levinson [27], who also gave an estimate of error in the approximate solution.

The other methods to obtain an asymptotic approximation or expansion of special functions and functions represented by integrations are stationary phase, Laplace, and steepest descent methods.

A high order of an algebraic or a differential equation or a large number of such equations are all manifestations of one of the principle difficulties that arise in solving physical problems. This difficulty is sometimes called *the imprecation of dimensionality*. The reduction of the dimensionality of a system is one of the approaches that one can take to deal with these sorts of differential equations. This method is also known as reduction in degrees of freedom, where one is able to carry out an asymptotic reduction of dimensionality. This way one can try to improve the solution obtained by using the asymptotic approximation. A typical example of such a situation is a three-body problem in elastic mechanics. The masses of celestial bodies (i.e., those of the Sun, the planet Jupiter, and the Earth), as a rule, differ noticeably, and a small parameter—the mass ratio—enable an asymptotic reduction of the dimensionality to be achieved. The classical methods of celestial mechanics are based on the limiting case to high symmetry assumption using an exactly solvable two-body problem.

It should be noted that asymptotic methods or the perturbation theory are often used without being specifically regarded as such and even without being fully understood. Thus, one-degree-of-freedom models are utilized extensively in engineering. Clearly, employing such models always involves an asymptotic reduction in the dimensionality and the possibility, at any rate in principle, of finding the corresponding corrections, but a clear indication that this is a rare case.

References

1. G.W. Bluman, J.D. Cole, *Similarity Methods for Differential Equations* (Springer, New York, 1943)
2. L. Dresner, *Similarity Solutions of Nonlinear Partial Differential Equations* (Pitman Advanced Publishing Program, 1983)
3. P.L. Sachdev, *Self-Similarity and Beyond: Exact Solutions of Nonlinear Problems* (Chapman & Hall/CRC, 2000)
4. G.I. Taylor, The formation of a blast wave by a very intense explosion, I, Theoretical discussion. Proc. R. Soc. A **201**, 159–174 (1950)
5. G.I. Taylor, The formation of a blast wave by a very intense explosion, II, The atomic explosion of 1945. Proc. R. Soc. A **201**, 175–186 (1950)
6. L. Sedov, *Similarity and Dimensional Methods in Mechanics* (Academic, New York, 1969)
7. G. Guderley, Starke kugelige und zylindrische Verdichtungsstosse in der Nahe des Kugelmittelpunktes bzw. der Zylinderachse. Luftfahrtforschung **19**, 302–312 (1942)
8. Y.B. Zel'dovich, Y.P. Raizer, *Physics of Shock Waves and High-Temperature Hydrodynamics Phenomena* (Dover Publication, New York, 2002)
9. G.I. Barenblatt, *Scaling, Self-similarity, and Intermediate Asymptotics* (Cambridge University Press, Cambridge, 1996)
10. G.I. Barenblatt, Y.B. Zel'dovich, Self-similar solutions as intermediate asymptotics. Annu. Rev. Fluid Mech. **4**, 295–312 (1972)
11. G.I. Barenblatt, *Scaling* (Cambridge University Press, Cambridge, 2003)
12. G. Birkhoff, *Hydrodynamics* (Princeton University Press, Princeton, NJ, 1950). Chap. V
13. L. Dresner, *On the Calculation of Similarity Solutions of Partial Differential Equations*, Oak Ridge National Laboratory Report ORNL/TM-7404 (1980)
14. R.M. Mattheij, S.W. Rienstra, J.H.M. ten Thije Boonkkamp, *Partial Differential Equations: Modeling, Analysis, Computation* (Society for Industrial & Applied Mathematics (SIAM), 2005)
15. G. Hansen, *Similarity Analysis of Boundary Value Problems in Engineering* (Prentice-Hall, New Jersey, 1964)
16. H. Schlichting, *Boundary Layer Theory* (McGraw-Hill, New York, 1960). Chap. VII
17. A.B. Vasilieva, V.F. Butuzov, *Asymptotical Methods in Theory of Singular Perturbation* (Vysshaia Shkola, Moscow, 1990). in Russian
18. J. Awrejcewics and V.A. Krysko, *Introduction to Asymptotic Methods* (Chapman & Hall/CRC, 2006)
19. A. Nuseir, *Symbolic Computation of Exact Solutions of Nonlinear Partial Differential Equations Using Direct Methods*, A thesis submitted to the Faculty and the Board of Trustees of the Colorado School of Mines in partial fulfillment of the requirements for the degree of Doctor of Philosophy (Mathematical and Computer Sciences)
20. P.A. Clarkson, M.D. Kruskal, New similarity reductions of the Boussinesq equation. J. Math. Phys. **30**, 2201–2213 (1989)
21. K.T. Vu, J. Bucher, J. Carminat, Similarity solutions of partial differential equations using DESOLV. Comput. Phys. Commun. **176**(11–12), 682–693 (2007)
22. K. Fakhar, Exact solutions for nonlinear partial differential equation arising in thirds grade fluid Flows. Southeast Asian Bull. Math. **32**, 65–70 (2008)

23. R.S. Rivlin, J.L. Ericksen, Stress deformation relations for isotropic materials. J. Ration. Mech. Anal. **4**, 323 (1955)
24. C. Truesdell, W. Noll, *The Nonlinear Field Theories of Mechanics*, 2nd edn. (Springer, New York, 1992)
25. C.M. Bender, S.A. Orszag, *Advanced mathematical Methods for Scientists and Engineers* (McGraw-Hill, New York, 1978)
26. G.I. Barenblatt, *Scaling Phenomena in Fluid Mechanics* (Cambridge University Press, New York, 1994)
27. N. Levinson, Asymptotic behavior of solution of nonlinear differential equations. Stud. Appl. Math. **49**, 285–297 (1969)

Appendix A: Simple Harmonic Motion

A.1 We Start with Hooke's Law

Harmonic oscillator is depicted here and Hooke's law defines the following equation:

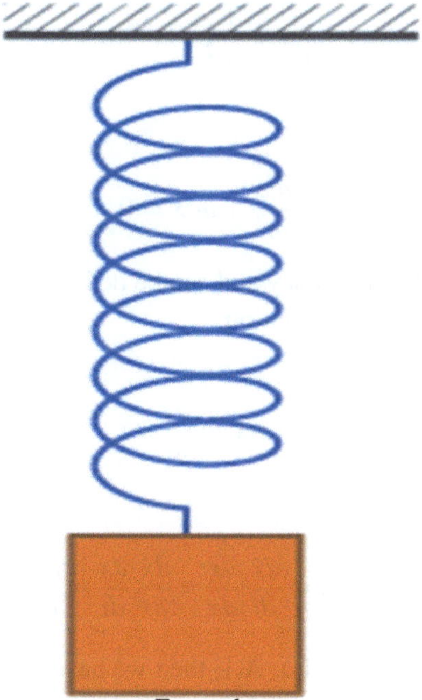

$$F = -kx.$$

But using Newton's second law of motion, we can write

B. Zohuri, *Dimensional Analysis Beyond the Pi Theorem*,
DOI 10.1007/978-3-319-45726-0

$$F = ma,$$

where in both equations F is the force, x is displacement of mass m, and k is spring constant as well as the acceleration of the mass.

By equating both equations, we obtain the following:

$$ma = -kx$$

or

$$m\frac{d^2x}{dt^2} = -kx$$

and

$$\frac{d^2x}{dt^2} + \frac{k}{m}x = 0.$$

Define $\omega_0^2 = \frac{k}{m}$ then we have

$$\frac{1}{\omega_0^2} = \frac{d^2x}{dt^2} + x = 0$$

or

$$\frac{d^2x}{dt^2} + \omega_0^2 x = 0 \tag{A.1}$$

where ω is called angular frequency and can be defined as $\omega = 2\pi f = \frac{2\pi}{T}$ where f is frequency and T is period of oscillation.

Define

$$\dot{x} = \frac{dx}{dt}$$

then we have

$$\frac{d^2x}{dt^2} = \ddot{x} = \frac{d\dot{x}}{dt}\frac{dx}{dx} = \frac{d\dot{x}}{dx}\frac{dx}{dt} = \frac{d\dot{x}}{dx}\dot{x}.$$

Substituting the above result in Eq. A.1, then we have

$$\frac{d\dot{x}}{dx}\dot{x} + \omega_0^2 x = 0,$$

$$\dot{x}\,d\dot{x} + \omega_0^2 x dx = 0.$$

Integrating over the differential equation, we have

$$\int \dot{x}\,d\dot{x} + \int \omega_0^2 x\,dx = 0,$$

$$\frac{1}{2}\dot{x}^2 + \frac{1}{2}\omega_0^2 x^2 = \text{cte},$$

$$\dot{x}^2 + \omega_0^2 x^2 = 2\text{cte} = K = (A\omega_0)^2,$$

$$\dot{x}^2 = A^2\omega_0^2 - \omega_0^2 x^2,$$

$$\frac{dx}{dt} = \dot{x} = \pm\omega_0\sqrt{A^2 - x^2},$$

Separating of variable gives the following results:

$$\frac{dx}{\sqrt{A^2 - x^2}} = \omega_0\,dt$$

or

$$\int \frac{dx}{\sqrt{A^2 - x^2}} = \int \omega_0\,dt. \tag{A.2}$$

Two possible solutions

$$\arcsin\frac{x}{A} = \omega_0 t + \phi,$$

$$\arccos\frac{x}{A} = \omega_0 t + \phi.$$

ϕ Integrating constant term

Note: To do left-hand side integral, we can do the following steps: Assume

$$x = A\ \sin y \Rightarrow dx = A\cos y dy$$

and

$$\sqrt{A^2 - x^2} = \sqrt{A^2 - A^2\sin^2 y} = A\sqrt{1 - \sin^2 y} = A\cos y.$$

Therefore,

$$\int \frac{dx}{\sqrt{A^2 - x^2}} = \int \frac{A\cos y}{A\cos y} dy = \int dy = y.$$

Since $x = A\sin y$, we can conclude the following:

$$\sin y = \frac{x}{A} \Rightarrow y = \arcsin\frac{x}{A}.$$

Therefore, if we substitute the last step result in Eq. A.2, then we have

$$\arcsin\frac{x}{A} = \int \omega_0 dt = \omega_0 t + \phi.$$

Similarly we can have same results for the second solution:

$$\arccos\frac{x}{A} = \int \omega_0 dt = \omega_0 t + \phi.$$

So the general solution is written as follows:

$$x = A\cos(\omega_0 t + \phi).$$

However, remember we assumed that $\omega = 2\pi f = \frac{2\pi}{T}$ where $f = \frac{1}{T}$ and f is frequency while T is the period.

Appendix B: Pendulum Problem

B.1 Definition

A pendulum is a mass (or bob) on the end of a string of negligible mass that, when initially displaced, will swing back and forth under the influence of gravity over its central (lowest) point. The regular motion of a pendulum can be used for time-keeping; pendulums are used to regulate (Fig. B.1).

A simple is an idealization, working on the assumption that:

- The rod or cord on which the bob swings is massless, inextensible, and always remains taut.
- The motion occurs in a two-dimensional plane, i.e., the bob does not trace an ellipse.
- The motion does not lose energy to friction.

The differential equation, which represents the motion of the pendulum very similar to simple harmonic motion, is

$$\frac{d^2\theta}{dt^2} + \frac{g}{l}\sin\theta = 0 \tag{B.1}$$

See Appendix A for Eq. B.1 derivation as well as the following pages.

In order to derive the simple pendulum equation and prove the dimensional analysis case about we show the following depiction (Fig. B.2):

Note: The path of the pendulum sweeps out an arc of a circle. The angle θ is measured in radians, and this is crucial for this formula. The blue arrow is the gravitational force acting on the bob, and the violet arrows are that same force resolved into components parallel and perpendicular to the bob's instantaneous motion. The direction of the bob's instantaneous velocity always points along the red axis, which is considered the tangential axis because its direction is always tangent to the circle. Consider Newton's second law:

B. Zohuri, *Dimensional Analysis Beyond the Pi Theorem*,
DOI 10.1007/978-3-319-45726-0

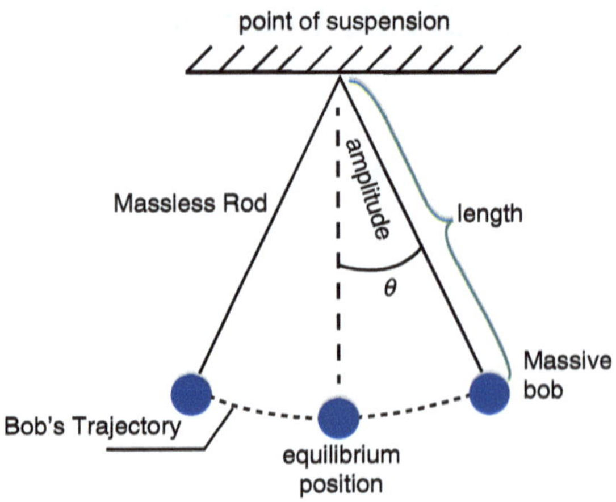

Fig. B.1 Simple gravity pendulum assumes no air resistance and no friction

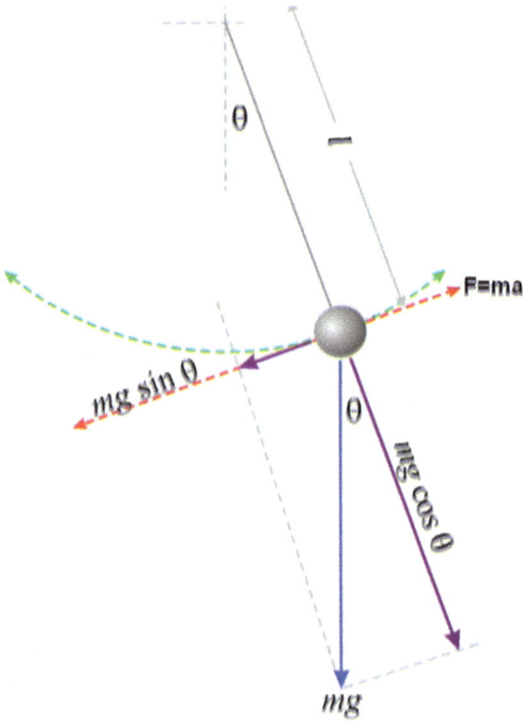

Fig. B.2 Force diagram of a simple gravity pendulum

$$F = ma,$$

where F is the sum of forces on the object, m is the mass, and a is the instantaneous acceleration. Because we are only concerned with changes in speed, and because the bob is forced to stay in a circular path, we apply Newton's equation to the tangential axis only. The short violet arrow represents the component of the gravitational force in the tangential axis, and trigonometry can be used to determine its magnitude. Thus,

$$\begin{aligned} F &= -mg\sin\theta = ma \\ a &= -g\sin\theta, \end{aligned}$$

where g is the acceleration due to gravity near the surface of the earth. The negative sign on the right-hand side implies that θ and a always point in opposite directions. This makes sense because when a pendulum swings further to the left, we would expect it to accelerate back toward the right.

This linear acceleration a along the red axis can be related to the change in angle θ by the arc length formulas; s is arc length:

$$\begin{aligned} s &= l\theta, \\ v &= \frac{ds}{dt} = l\frac{d\theta}{dt}, \\ a &= \frac{d^2s}{dt^2} = l\frac{d^2\theta}{dt^2}. \end{aligned}$$

Thus,

$$l\frac{d^2\theta}{dt^2} = -g\sin\theta$$

or

$$\frac{d^2\theta}{dt^2} + \frac{g}{l}\sin = 0, \tag{B.2}$$

This is the differential equation which, when solved for $\theta(t)$, will yield the motion of the pendulum. It can also be obtained via the conservation of mechanical energy principle: any given object, which fell a vertical distance h, would have acquired kinetic energy equal to that which it lost to the fall. In other words, gravitational potential energy is converted into kinetic energy. Change in potential energy is given by

$$\Delta U = mgh$$

change in kinetic energy (body started from rest) is given by

$$\Delta K = \frac{1}{2}mv^2.$$

Since no energy is lost, those two must be equal:

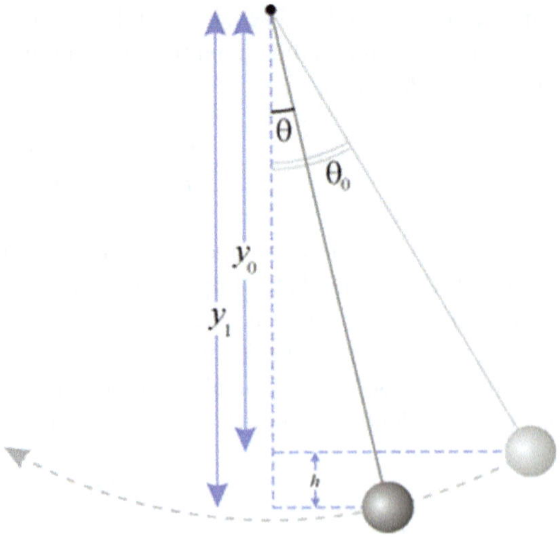

Fig. B.3 Trigonometry of a simple gravity pendulum

$$\frac{1}{2}mv^2 = mgh,$$
$$v = \sqrt{2gh}.$$

Using the arc length formula above, this equation can be rewritten in favor of $\frac{d\theta}{dt}$

$$\frac{d\theta}{dt} = \frac{1}{l}\sqrt{2gh},$$

where h is the vertical distance the pendulum fell. Consider Fig. B.3. If the pendulum starts its swing from some initial angle θ_0, then y_0, the vertical distance from the screw, is given by

$$y_0 = l\cos\theta_0$$

similarly, for y_1, we have

$$y_1 = l\cos\theta$$

then h is the difference of the two

$$h = l(\cos\theta - \cos\theta_0)$$

substituting this into the equation for $\frac{d\theta}{dt}$ gives

$$\frac{d\theta}{dt} = \sqrt{\frac{2g}{l}(\cos\theta - \cos\theta_0)}. \tag{B.3}$$

This equation is known as the *first integral of motion*; it gives the velocity in terms of the location and includes an integration constant related to the initial displacement (θ_0). We can differentiate, by applying the chain rule, with respect to time to get the acceleration:

$$\begin{aligned}
\frac{d}{dt}\frac{d\theta}{dt} &= \frac{d}{dt}\sqrt{\frac{2g}{l}(\cos\theta - \cos\theta_0)} \\
\frac{d^2\theta}{dt^2} &= \frac{1}{2}\frac{-(2g/l)\sin\theta}{\sqrt{(2g/l)(\cos\theta - \cos\theta_0)}}\frac{d\theta}{dt} \\
&= \frac{1}{2}\frac{-(2g/l)\sin\theta}{\sqrt{(2g/l)(\cos\theta - \cos\theta_0)}}\sqrt{\frac{2g}{l}(\cos\theta - \cos\theta_0)} = -\frac{g}{l}\sin\theta \\
\frac{d^2\theta}{dt^2} &= -\frac{g}{l}\sin\theta,
\end{aligned}$$

which is the same result as obtained through force and dimensional analysis.

Appendix C: Similarity Solution Methods for Partial Differential Equations (PDEs)

Here we discuss briefly how to handle and solve a partial differential equation of high order by reducing to and ordinary differential equation using self-similar methods given by George W. Bluman and J. D. Cole.

C.1 Self-Similar Solutions by Dimensional Analysis

Consider the diffusion problem from the last section, with point-wise release (reference: *Similarity Methods for Differential Equations* (Applied Mathematical Sciences, Vol. 13)—Paperback (Dec. 2, 1974) by George W. Bluman and J.D. Cole (Sect. 2.3):

$$\begin{cases} \dfrac{\partial c}{\partial t} = D\dfrac{\partial^2 c}{\partial x^2} + Q_0\delta(x)\delta(t) \\ c(x,0) = 0, \quad c(\pm\infty, t) = 0. \end{cases}$$

Initial release within infinitely narrow neighborhood of $x = 0$, such that $\Pi(x)/d = \delta(x)$ and $L/d \to \infty$. Note Q_0 has different dimension as the previous Q because of the cross-sectional area S and time contained in $\delta(t)$.

1. *Dimensional analysis*
   $\{c\} = ML^{-3}, \{D\} = L^2T^{-1}, \{Q_0\} = ML^{-2}$ (mass release per unit cross-sectional area) $\{x\} = L$, $\{t\} = T$. Thus, we expect 2Pi groups:

$$\Pi_1 = \frac{\sqrt{Dt}}{Q_0}c, \quad \Pi_2 = \frac{x}{\sqrt{Dt}}$$

   and the solution to the PDE problem must be of the form $\Pi_1 = f(\Pi_2)$ or

B. Zohuri, *Dimensional Analysis Beyond the Pi Theorem*,
DOI 10.1007/978-3-319-45726-0

$$c = \frac{Q_0}{\sqrt{Dt}} f\left(\frac{x}{\sqrt{Dt}}\right).$$

Normally we expect dimensional analysis to reduce the number of variables and parameters. However, here we reduce the number of independent variables from 1 to 1!

2. *Transformation of PDE to ODE*
   Now we can plug this form back into the PDE. First, the partial derivatives:

$$\frac{\partial c}{\partial t} = -\frac{Q_0}{2t\sqrt{Dt}} f - \frac{Q_0 x}{2Dt^2} f', \quad \frac{\partial c}{\partial x} = \frac{Q_0}{Dt} f', \quad \frac{\partial^2 c}{\partial x^2} = \frac{Q_o}{(Dt)^{3/2}} f''.$$

For $t > 0$, there is no more injection: $\delta(t) = 0$. After inserting the above into the PDE:

$$-\frac{f}{2} - \frac{x}{2\sqrt{Dt}} f' = f'' \quad \text{or} \quad f'' + \frac{\xi}{2} f' + \frac{f}{2} = 0, \tag{C.1}$$

where $\xi = \frac{x}{\sqrt{Dt}}$ is our new independent variable. We have successfully transformed the PDE into an ODE. How about the initial and boundary conditions? Note that $t = 0$ and $x = \infty$ both correspond to $\xi = \infty$, so that the initial and boundary conditions can be rolled into one:

$$f(\pm\infty). \tag{C.2}$$

However, we need another condition on *f*, one that reflects the amount of initial injection. This is obtained by integrating the PDE over the following intervals:

$$\int_{0^-}^{t} dt \int_{-\infty}^{+\infty} [\text{PDE}] dx, \qquad \text{where } t = 0^- \text{ means "just before } t = 0\text{"}.$$

Now the left-hand side is

$$\int_{0^-}^{t} dt \int_{-\infty}^{+\infty} \frac{\partial^2 c}{\partial x^2} dx = \int_{-\infty}^{+\infty} dx \int_{0^-}^{t} \frac{\partial c}{\partial t} dt = \int_{-\infty}^{+\infty} [c(x,t) - c(x,0)] dx = \int_{-\infty}^{+\infty} c(x,t) dx.$$

Now we have $\int_{-\infty}^{+\infty} c(x,t) dx = Q_0$, which can be transformed, using the variable ξ, into

$$\int_{-\infty}^{+\infty} f(\xi) d\xi = 1. \tag{C.3}$$

ODE Eq. C.1, along with condition Eqs. C.2 and C.3, will uniquely determine $f(\xi)$, from which we get $c(x, t)$. We are not concerned with the actual solution of the new ODE problem. Rather, the interesting question is *how did we manage to turn a PDE to an ODE*.

3. *Discussion*

    (a) The problem admits a *self-similar solution:* if x is scaled by the diffusion length $(Dt)^{1/2}$, then the $c(x, t)$ profiles at different times can be collapsed onto each other if c is scaled by $Q_0/(Dt)^{1/2}$
    (b) This means that x and t are not really two independent variables; as far as c is concerned, they can be rolled into one independent variable ξ.
    (c) Similarity solutions are "happy coincidences" in physical process. Can we always find them for any PDEs? No. This problem is special in that there is no inherent length scale. Thus, we are not able to form dimensionless groups for each of the variables x, t and c; instead, we have to combine them and end up with only 2Pi groups. That is how we ended up with ODE. If we had the release length dS or the domain length L, the self-similar will be ruined.
    (d) Can we always find similarity solutions by dimensional analysis? No. However, we will study another example next and then introduce the general "stretching transformation" idea for detecting similarity solutions.

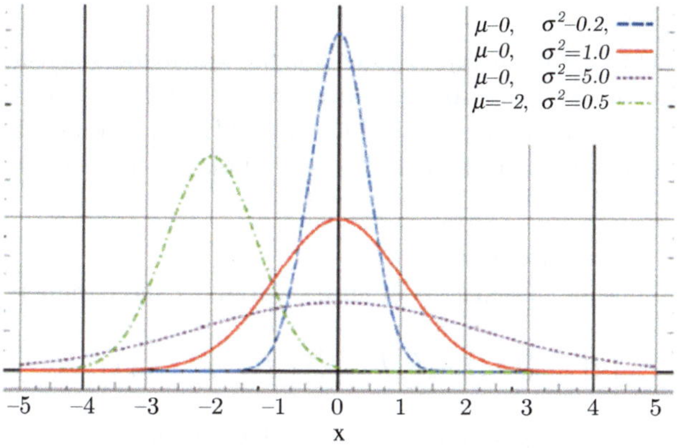

C.2 Similarity Solutions by Stretching Transformation

It is rare that similarity solutions can be obtained from dimensional analysis. In this section, we introduce the idea of stretching transformation which is a more general procedure for seeking out similarity in PDE problems. The materials are based on Barenblatt (Sect. 5.2) and Bluman and Cole (Sect. 2.5).

As a concrete example, we will take Prandtl's boundary-layer equation for flow over a flat semi-plane. After the boundary-layer approximation (that viscosity acts only within a thin layer, that the gradient in the flow direction (x) is much smaller than in the transverse direction (y), and that the pressure is constant in the y direction), the governing equations are

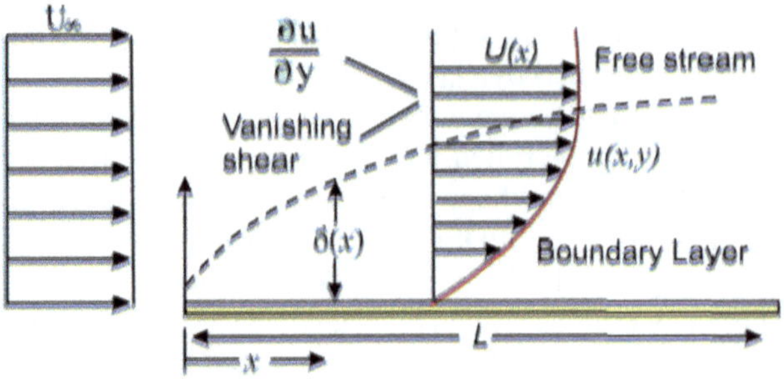

$$\begin{cases} u\dfrac{\partial u}{\partial x} + \upsilon\dfrac{\partial u}{\partial y} = \nu\dfrac{\partial^2 u}{\partial y^2} \\ \dfrac{\partial u}{\partial x} + \dfrac{\partial \upsilon}{\partial y} = 0 \\ u(x,0) = 0, \quad \upsilon(x,0) = 0 \\ u(x,\infty) = U_\infty, \quad u(0,y) = U_\infty \end{cases}$$

where U_∞ is the free-stream velocity and ν is the kinematic viscosity. If you recall your fluid mechanics, this problem does have a similarity solution (Blasius's solution), and the PDE can be reduced to ODE. (Try to distinguish the velocity υ from the viscosity ν. We could use different symbols but these are the conventional ones).

1. *Would dimensional analysis work?*
   Let us write out the dimensions of all the variables and parameters:

$$\{u\} = \{\upsilon\} = \{U_\infty\} = L/T, \quad \{\nu\} = L^2/T, \quad \{x\} = \{y\} = L$$

   There are two independent dimensions involved (L and T), and we can construct four

$$\Pi_1 = \frac{u}{U_\infty}, \quad \Pi_2 = \frac{\upsilon}{U_\infty}, \quad \Pi_3 = \frac{U_\infty x}{\nu}, \quad \Pi_4 = \frac{U_\infty y}{\nu}$$

   in addition, and we expect solutions such as

$$\Pi_1 = f(\Pi_3, \Pi_4), \quad \Pi_2 = g(\Pi_3, \Pi_4).$$

Plugging these back into the equations, we will see that we have **not** achieved a reduction of the number of independent variable. Dimensional analysis has failed to give us the similarity solution. Why? Even through the problem has no intrinsic time or length scales. There are only two indecent dimensions (L and T) instead of three. Thus, it is possible for x and y to form their own Pi groups; they do not have to be forced into a single one.
It turns out that in this particular example, a trivial manipulation can "cure" the above problem. This is not a general technique, but nevertheless, it is fun to illustrate here. We will take this little detour before marching into the general technique that is the focus of this section. Based on the physical insight that things happen at different scales along the x and y directions, which is the fundamental idea behind the boundary-layer approximation, we assign two different dimensions to x and y, L and H, and for the moment pretend that they are different dimensions. Now the list of variables and unknowns are scaled as such:

$$\{u\} = \{UÀ\hat{U}\} = L/T, \quad \{v\} = H/T, \quad \{£h\} = H2/T, \quad \{x\} = L, \quad \{y\} = H.$$

There are now three independent dimensions involved (L, H, and T), and we can construct only three dimensionless groups out of these:

$$\tilde{\Pi}_1 = \frac{u}{U_\infty}, \quad \tilde{\Pi}_2 = \sqrt{\frac{\upsilon}{\upsilon U_\infty / x}}, \quad \tilde{\Pi}_3 = \frac{y}{\sqrt{\nu x / V}} = \zeta.$$

Now we expect a similarity solution in this form:

$$u = U_\infty f(\zeta), \quad \upsilon = \sqrt{\frac{\nu U_\infty}{x}} g(\zeta).$$

Plugging this into the original PDE will show that, indeed, we have reduced the PDE problem to a couple of ODEs, whose solution is detailed in *Fluid Mechanics* textbooks. For another example of such "ingenious" dimensional analysis, see the Rayleigh problem analyzed in the next section (see also Bluman and Cole, p. 195). We typically seek to increase the number of independent dimensions (as done above) or decrease the number of dimensional parameters (as done in Bluman and Cole's example).

2. *Stretching transformation*
The "ingenious" dimensional analysis method is specific to the problems. There is, however, a general scheme for seeking out possible similarity solutions. The scheme sometimes goes by the name of "renormalization groups" or "invariant transformation groups" and is based on rather formalistic mathematical manipulations. We will skip the proofs and focus on the technique itself.
Since the essence of similarity is that the solution is *invariant* after certain scaling of the independent and dependent variables, we consider the following

stretching transformation, and see if such transformations will leave the PDE and the boundary conditions invariant.
Consider:

$$\begin{cases} U = \alpha^a u, & V = \alpha^b v \\ X = \alpha^c x, & Y = \alpha^d y \end{cases},$$

where α is a positive number. Under this transformation, we have

$$\frac{\partial u}{\partial x} = \alpha^{c-a}\frac{\partial U}{\partial X}, \quad \frac{\partial u}{\partial y} = \alpha^{d-a}\frac{\partial U}{\partial Y}, \quad \frac{\partial v}{\partial y} = \alpha^{d-a}\frac{\partial V}{\partial Y}, \quad \frac{\partial^2 u}{\partial y^2} = \alpha^{2d-a}\frac{\partial^2 U}{\partial Y^2}.$$

Plugging these into the original PDE and boundary conditions, we will see what choices of a, b, c, and d may maintain the invariance of the problem. The continuity equation yields:

$$c - a = d - b.$$

The three terms of the momentum equation requires:

$$c - 2a = d - a - b = 2d - a.$$

Note that the first equation above is identical to the preceding equation, and thus the momentum equation adds only one additional constraint on the power indices. Finally the boundary conditions require

$$a = 0$$

because for the problem in the new variables to be invariant, the nonhomogeneous BC should remain as $U(X_\infty) = U_\infty$. Now we have three equations that constrain the four indices, and we rewrite the transformation as

$$\begin{cases} U = u, & V = \dfrac{v}{\varepsilon} \\ X = \varepsilon^2 x, & Y = \varepsilon y \end{cases}, \quad \text{where} \quad \varepsilon = \alpha^d.$$

This transformation will leave the problem the same as before, in the new "stretched" and scaled variables. The fact that this *one-parameter family* of transformations will maintain the invariance of the PDE problem reveals the intrinsic self-similarity of the problem. In other words, if we stretch the coordinate y by a factor ε, then we must stretch x by ε^2 and the velocity component ε^2 by $1/\varepsilon$ in order to collapse the velocity profiles. From this argument, we recognize that

$$u, \upsilon\sqrt{x}, \frac{y}{\sqrt{x}}$$

shall remain the same no matter how we stretch the coordinates. These are known as the invariants of the transformation, and immediately suggest the following similarity solution:

$$\begin{cases} u = f(\zeta) \\ \upsilon = \frac{1}{\sqrt{x}} g(\zeta) \end{cases}, \quad \text{with the similarity variable } \zeta = \frac{y}{\sqrt{x}}.$$

This is the same form as obtained from the "ingenious dimensional analysis," aside from a few constant factors. Note that we reached the conclusion here not through dimensional considerations, but through the idea of invariance under general stretching transformations.

Now it is a simple matter to plug these forms into the original PDE problem, and transform it into the following ODE problem:

$$\begin{cases} \upsilon f'' + f'\left(\frac{\zeta}{2} f - g\right) = 0 \\ \zeta f' - 2g' = 0 \\ f(\infty) = U_\infty, f(0) = 0,\ g(0) = 0 \end{cases}$$

the solution of which will not be of immediate interest to us here. Note that the two BCs at $x = 0$ and $y = \infty$ both project onto $\zeta = \infty$.

C.3 Similarity Solution for the Rayleigh Problem

The Rayleigh problem is another classical example with a self-similar solution. Consider the transient motion in a viscous fluid induced by a flat plate moving in its own plane. Initially both the plate and the fluid are at rest. Starting at, the plate moves with a constant velocity. The Nervier–Stokes equations, simplified for this problem, along with the initial and boundary conditions, can be written as

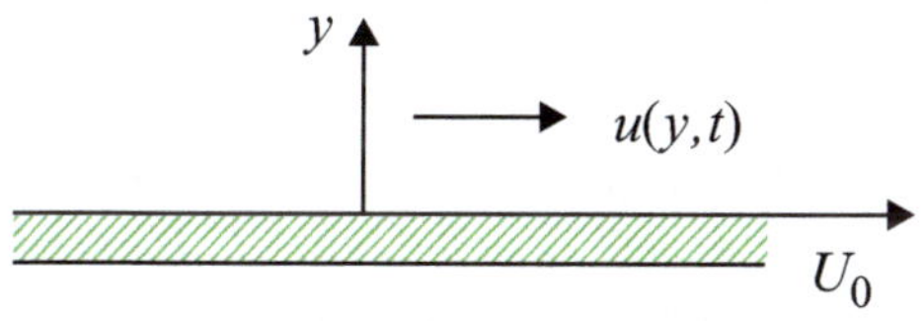

$$\begin{cases} \dfrac{\partial u}{\partial t} = \nu \dfrac{\partial^2 u}{\partial y^2} \\ u(y,0) = 0, \quad u(0,t) = U_0, \quad u(\infty, t) \end{cases}.$$

(a) *Dimensional analysis*
From the following dimensions,

$$\{u\} = \{U_0\} = L/T, \quad \{\nu\} = H^2/T, \quad \{t\} = T, \quad \{y\} = L,$$

where ν is fluid viscosity.
We can make 3Pi groups, say u/U_0, $U_0 y/\nu$, and $U_0^2 t/\nu$, and there is no reduction to ODE. Again, we can play trick here, by either *increasing* the number of "independent dimension" or *decreasing* the number of parameters, so as to reduce the number of Pi groups.
Using the physical observation that viscous diffusion happens along the y direction, while the primary flow is in the x direction, we can introduce different length scales:

$$\{u\} = \{U_0\} = L/T, \quad \{\nu\} = H^2/T, \quad \{t\} = T, \quad \{y\} = H.$$

Now there are only 2Pi groups:

$$\Pi_1 = \frac{u}{U_0}, \quad \Pi_2 = \frac{y}{\sqrt{\nu t}}$$

in addition, we can try a similarity solution of the form

$$u(y,t) = U_0 f\left(\frac{y}{\sqrt{\nu t}}\right)$$

Alternatively, we can reduce the number of parameters by scaling u by U_0, and calling $\tilde{u}(y,t) = u(x,t)/U_0$ the new dependent variable. Now the problem has one less parameter, and again only admits 2Pi groups. In the following, however, let us carry out the formal procedure of *stretching transformation* as an exercise.

(b) *Stretching transformation*
Consider

$$U = \alpha^a u, \quad Y = \alpha^b, \quad T = \alpha^c t,$$

where α is a positive number. Under this transformation, we have

$$\frac{\partial u}{\partial t} = \alpha^{c-a}\frac{\partial U}{\partial T}, \qquad \frac{\partial^2 u}{\partial y^2} = \alpha^{2b-a}\frac{\partial^2 U}{\partial Y^2}.$$

To maintain invariance of the PDE, we require

$$c - a = 2b - a \quad \text{or} \quad c = 2b.$$

The boundary condition $u(0, t) = U_0$ requires $a = 0$. Thus, the following transformation renders the problem invariant:

$$U = u, \quad Y = \varepsilon y, \quad T = \varepsilon^2 t, \quad \text{which} \quad \varepsilon = \alpha^b.$$

This transformation dictates that y and t be transformed in a coordinated way. Thus u and $\zeta = y/\sqrt{t}$ shall be our new variables that remain unchanged for any stretching α or ε:

$$u = f\left(\frac{y}{\sqrt{t}}\right) = f(\zeta)$$

This reduces the original PDE into the following ODE problem:

$$\begin{cases} 2\nu f'' + \zeta f' = 0 \\ f(0) = U_\infty, \quad f(\infty) = 0 \end{cases},$$

which can be integrated analytically to give:

$$f = c_1 \int_0^\zeta \exp\left(-\frac{z^2}{4\nu}\right) dz + c_2.$$

Noting that $\int_0^\zeta \exp\left(-\frac{z^2}{4\nu}\right) dz = 2\sqrt{\nu}\int_0^\infty \exp(-\xi^2)d\xi = \sqrt{\pi\nu}$, the two constants of integration are determined:

$$c_1 = -U_0/\sqrt{\pi\nu}.$$
$$c_2 = U_0$$

Finally, the solution can be written in terms of the *complementary error function*:

$$f = U_0 \text{erfc}\left(\frac{\zeta}{2\sqrt{\nu}}\right) = U_0 \text{erf}\left(\frac{y}{\sqrt{4\nu t}}\right)$$

with $\text{erfc}(x) = 1 - \frac{2}{\sqrt{\pi}}\int_0^x \exp(-z^2) dz$.

Index

B. Zohuri, *Dimensional Analysis Beyond the Pi Theorem*,
DOI 10.1007/978-3-319-45726-0

MIX
Papier aus verantwortungsvollen Quellen
Paper from responsible sources
FSC® C105338
FSC
www.fsc.org

If you have any concerns about our products,
you can contact us on
ProductSafety@springernature.com

In case Publisher is established outside the EU,
the EU authorized representative is:
Springer Nature Customer Service Center GmbH
Europaplatz 3, 69115 Heidelberg, Germany

Printed by Libri Plureos GmbH
in Hamburg, Germany